KAWASAKI

KAWASAKI

MICK WALKER

OSPREY
AUTOMOTIVE

Published in Great Britain in 1993 by Osprey, an imprint of Reed Consumer Books Limited, Michelin House, 81 Fulham Road, London SW3 6RB and Auckland, Melbourne, Singapore and Toronto.
Reprinted spring 1997

© Reed International Books 1993, 1997

All rights reserved. Apart from any fair dealing for the purposes of private study, research, criticism or review, as permitted under the Copyright, Designs and Patents Act, 1988, no part of this publication may be reproduced, stored in a retrieval system, or transmitted in any form or by any means electronic, electrical, chemical, mechanical, optical, photocopying, recording or otherwise, without prior written permission.
All enquiries should be addressed to the Publisher.

ISBN 1 85532 675 2

Editor Shaun Barrington
Page design Paul Kime

Printed in Hong Kong

Front cover
ZXR 750, called in 1996 the ZXR 7R, was first introduced in 1989. The Ninja was a liquid cooled 16-valve in-line four, 125bhp, and weighed 200kg. The bhp has increased through the years, and in 1996 you could get a ZXR 1100. (Garry Stuart)

Back cover
The Cyclone is a special produced by Daytona, a dealer in Middlesex, England, based upon the Zephyr 1100. Air-cooled, 1062cc, the machine has various modifications including to camshafts and cylinder heads, plus Olin shocks, 43mm stiffened forks, Harrison brake calipers, and 17-in (not 18) soft compound Michelin at the front. (Roland Brown)

Half-title page
Kurt Nicoll, runner-up in the 1989 500 cc World Motocross series

Title page
The smaller ZZ-R was Kawasaki's first 600 cc class machine to feature an aluminium main frame. The engine, although based on the GPX, had been strengthened and given a power boost to almost 100 bhp

Illustrations by Don Morley or from Mick Walker's collection

For a catalogue of all books published by Osprey Automotive please write to:
The Marketing Department, Reed Consumer Books,
1st Floor, Michelin House, 81 Fulham Road, London SW3 6RB

Contents

Introduction 6
Early days 10
Lightweights 18
Air-cooled fours 26
Trail bikes 48
Racing 56
Four-stroke twins 68
Z1300 76
Custom cruisers 82
Off-road competition 90
GTR 94
Liquid-cooled fours 102
Two-stroke twins 124

Future world champion Eddie Lawson campaigned this 'sit-up-and-beg' Z1000R in American racing with considerable success during the early 1980s

Introduction

Kawasaki is very much an engineer's company. Engineers were strongly represented in the senior management of the company established in 1878. They still are today – even though Kawasaki Heavy Industries now comprise no less than seven major divisions, just one of which is responsible for the design and production of engines (including diesel and gasoline types) and motorcycles.

The Kawasaki portfolio also consists of Aircraft, Energy Plant Engineering, Machinery, Plant Engineering, Rolling Stock and Ships.

With these diverse operations under a single umbrella, Kawasaki Heavy Industries has built a vast store of engineering experience and new technology. One of the largest R & D programmes in the world ensures that all divisions benefit from such new engineering advances as

Above
Smallest of the air-cooled Kawasaki fours at the time, the 1980 Z400J1

Right
The all-new ZXR750 made its debut in H1 form for the 1989 season

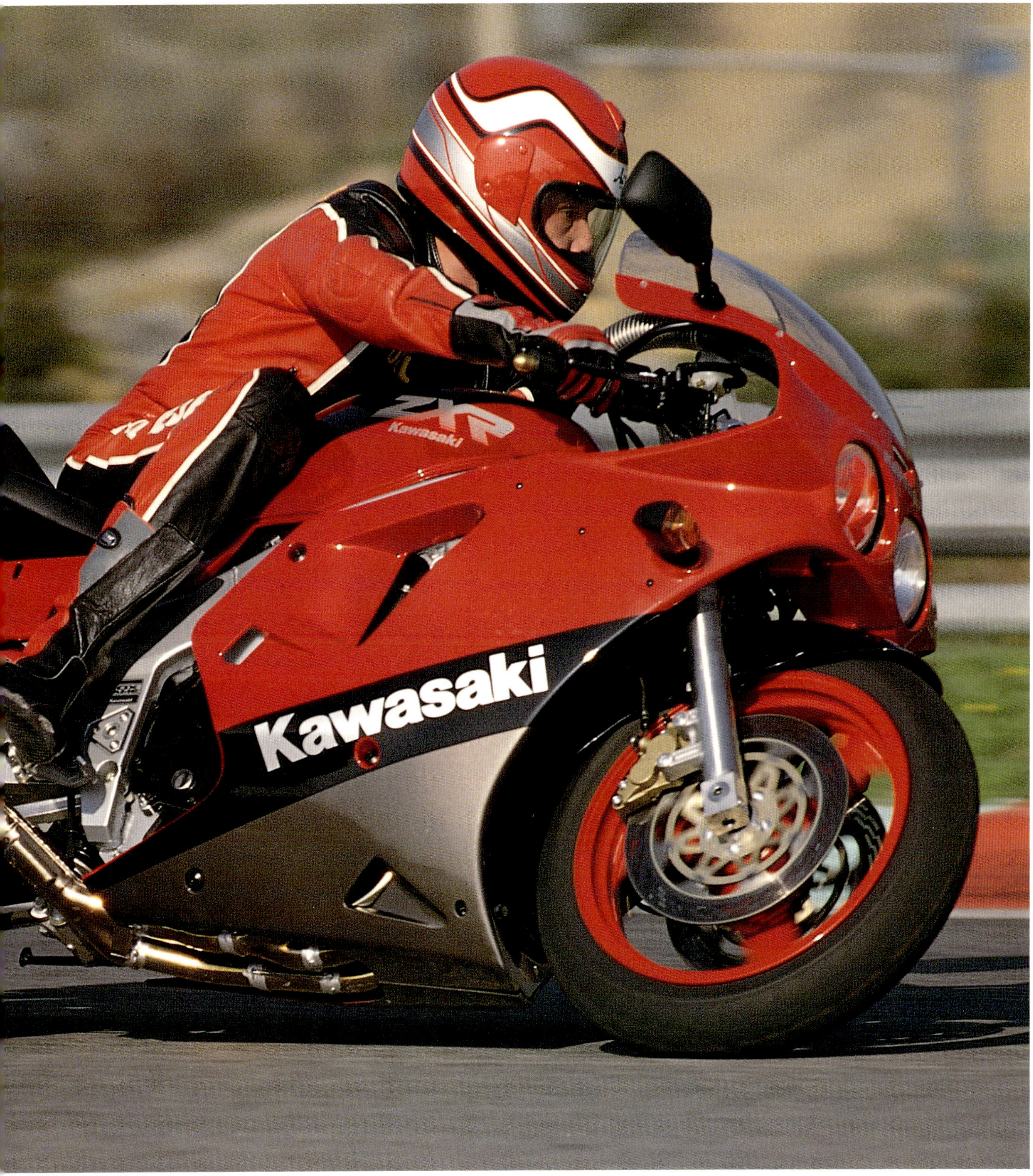
ZXR
Kawasaki
Kawasaki

World Champion Wayne Rainey had many of his early races aboard an air-cooled four-cylinder Kawasaki in the early 1980s

integrated CAD/CAM techniques, while constant technological exchange between its divisions helps Kawasaki maintain its engineering prowess.

For example, aircraft research ensures that all divisions have the latest data on lightweight, high-strength metal alloys and synthetic materials; R & D on such varied powerplants as massive oil tanker engines (producing up to 4000 horsepower per cylinder!) and sub-compact automobile engines ensures the widest range of technology on every aspect of internal combustion; research on computer-control systems for aircraft, ships and plants provides up-to-the-minute electronics technology; and production of industrial robots allows maximum production efficiency and quality control.

With an emphasis on high performance and quality, the Motorcycle Division has contributed to this process by developing a complete range of two-wheelers over the last three decades. For the last 20 years Kawasaki has created a number of world-class, trend setting bikes including the Mach III, Z1, GPZ900R, ZZ-R1100 and the ZXR750 Superbike racer.

Like other Kawasaki Heavy Industries divisions, the bike building operation is centred west of Oskaka, separated geographically from the

GPZ 1000RX was short-lived, replaced pretty swiftly by the ZX10 and ultimately the superb ZZ-R1100

other Japanese motorcycle manufacturers. The company has also begun the production of certain models in other countries; notably the ZZ-R1100, which is now built in an American factory. In this respect Kawasaki are following a lead set by Honda with their top-of-the-range Gold Wing model. And like Honda, Kawasaki see their future very much as an international one, not just confined to a Japanese manufacturing base. This just has to be a good thing for both the future of motorcycling and more broadly for the future of world trade as the new century beckons.

Early days

Unlike its other Japanese rivals, Kawasaki can rightly claim that motorcycling is only a small part of its vast empire, because Kawasaki Heavy Industries is one of the world's largest industrial complexes with interests in shipping, aviation and locomotion, among others. It can trace its origins back to the year 1878 when Shozo Kawasaki founded a shipyard at Tsukiji, Tokyo. In 1881, the Kawasaki Hyogo Shipyard was formed at Higashide-cho, Hyogo, and five years on, the two yards were combined to form the Kawasaki Shipyard. A decade later, in 1896, this became incorporated and Kojiro Matsukata was made president. Thereafter, it grew rapidly into one of Japan's biggest industrial organisations.

The first locomotive appeared in 1901, and in 1911, Kawasaki entered the field of marine transportation. During World War 1, more muscle was put on with a major new plant to manufacture steel. Soon after this, automobile and aviation divisions were established, the latter including production of their own engines. During the 1920s and 1930s, each division became prominent in its own sphere as a leader in Japanese industry. For example, many of the country's finest aircraft were of Kawasaki's design and manufacture, and the major customer was the fast-expanding Japanese Army Air Force. When war came again, Kawasaki was one of the largest suppliers to the military authorities, and the Ki-61 *Hein* was the only Japanese fighter aircraft of the conflict with a liquid-cooled engine.

When the war ended, Kawasaki, unlike many other companies, found its engineering skill in such demand that only one of its many plants was standing idle. This, too, soon found a market niche to keep it busy – producing engine and gearbox assemblies for the rapidly emerging motorcycle industry. One of Kawasaki's customers was the old established Meguro concern, and it was with this company that Kawasaki took its first steps towards becoming a motorcycle manufacturer in its own right.

The move could be traced back to 1959 when a motorcycle research and development centre was set up. From this came a new assembly plant in Akashi, which was completed in 1960, and late the same year Meguro became affiliated to Kawasaki Aircraft. This link was strengthened in 1961 when there was a further move to bond Kawasaki and Meguro into one corporate body, with its head quarters in Tokyo. In mid 1961, Kawasaki Auto Sales was formed, and then in 1962 came the first model to carry a Kawasaki badge. This was the 125 cc B8, a single-cylinder two-stroke.

The same year, Meguro changed its name to Kawasaki Meguro and the

Above right
The 646 cc (66 × 72.6 mm) W1SS twin of 1968 produced 50 bhp at the crankshaft and was a development of the W1 which had been introduced two years earlier. Both were derived from the older Meguro models, which were themselves taken from the BSA A7/10 series

Right
The very first version of the 500 three-cylinder Kawasaki, the Mach III, made its debut in 1969. Here Dutchman Henk Vink is putting one through its paces in March of that year. Vink covered the 400 metre distance in 13.8 seconds, a terminal velocity of 104.82 mph

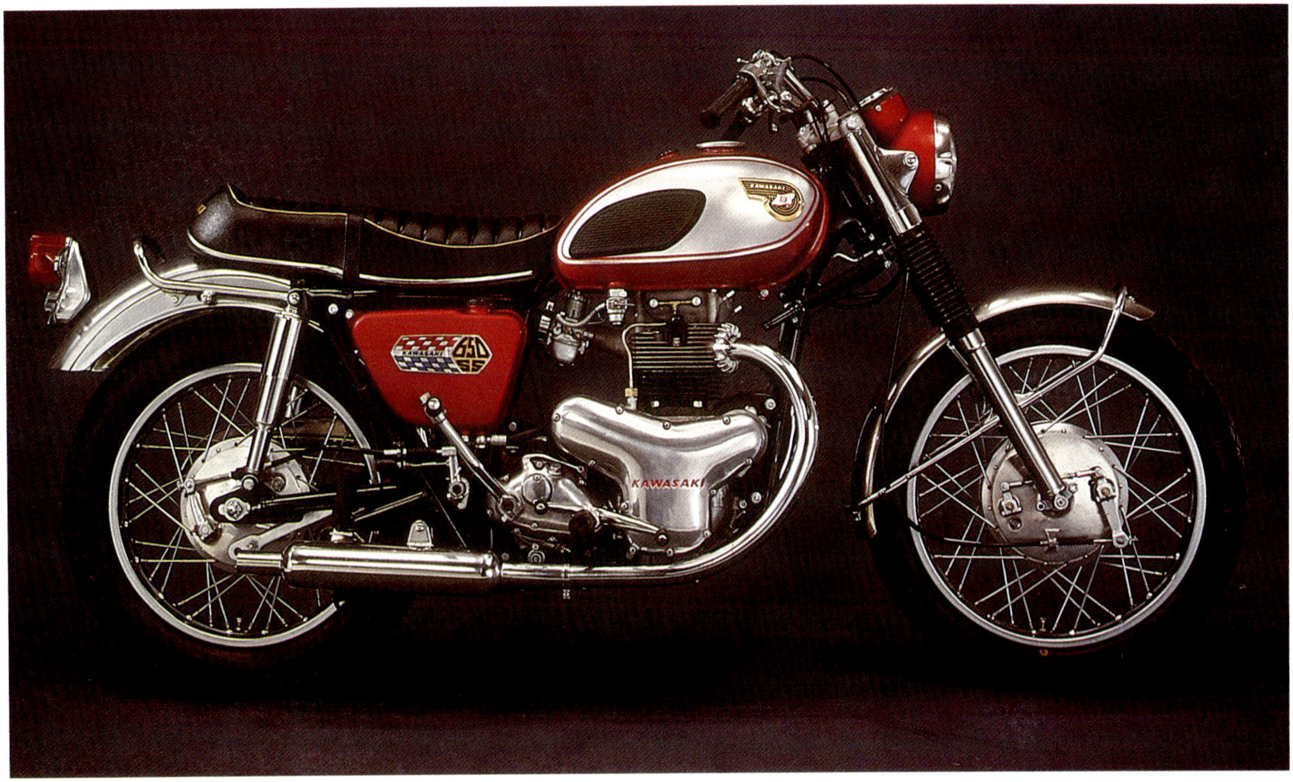
KAWASAKI

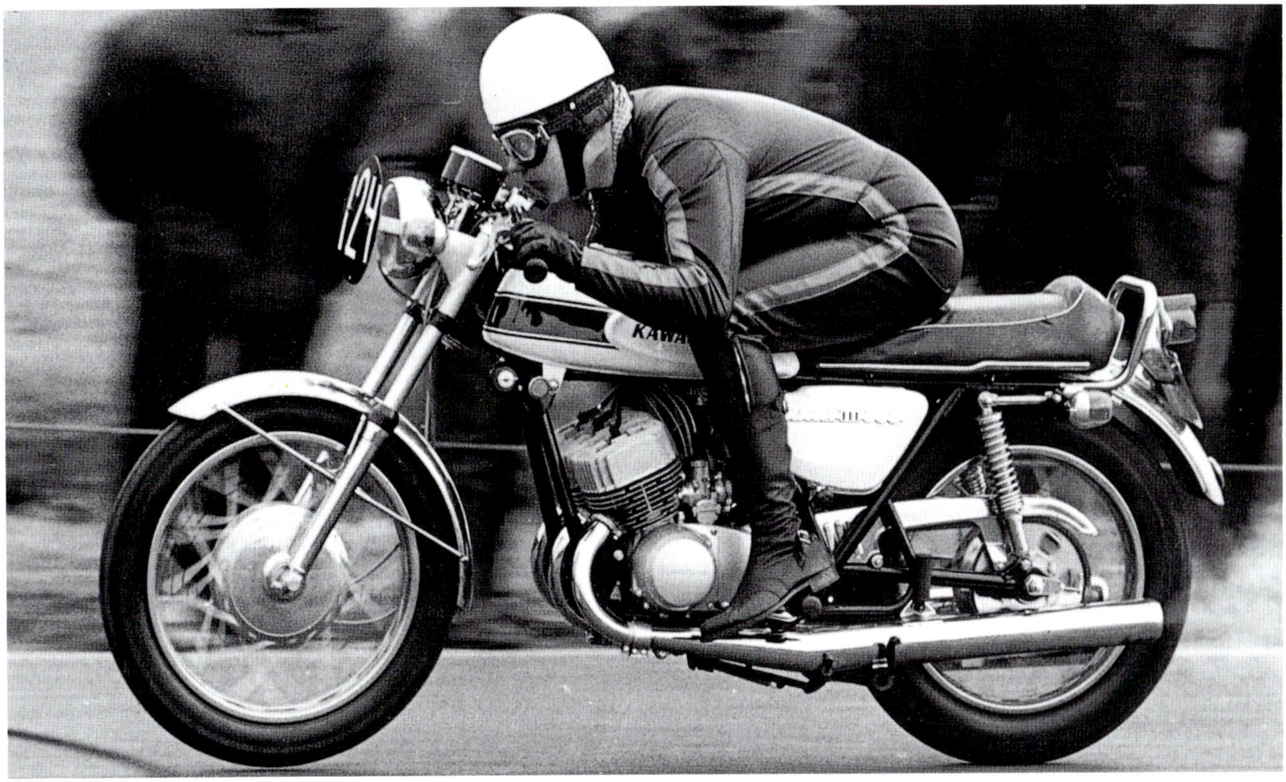
124

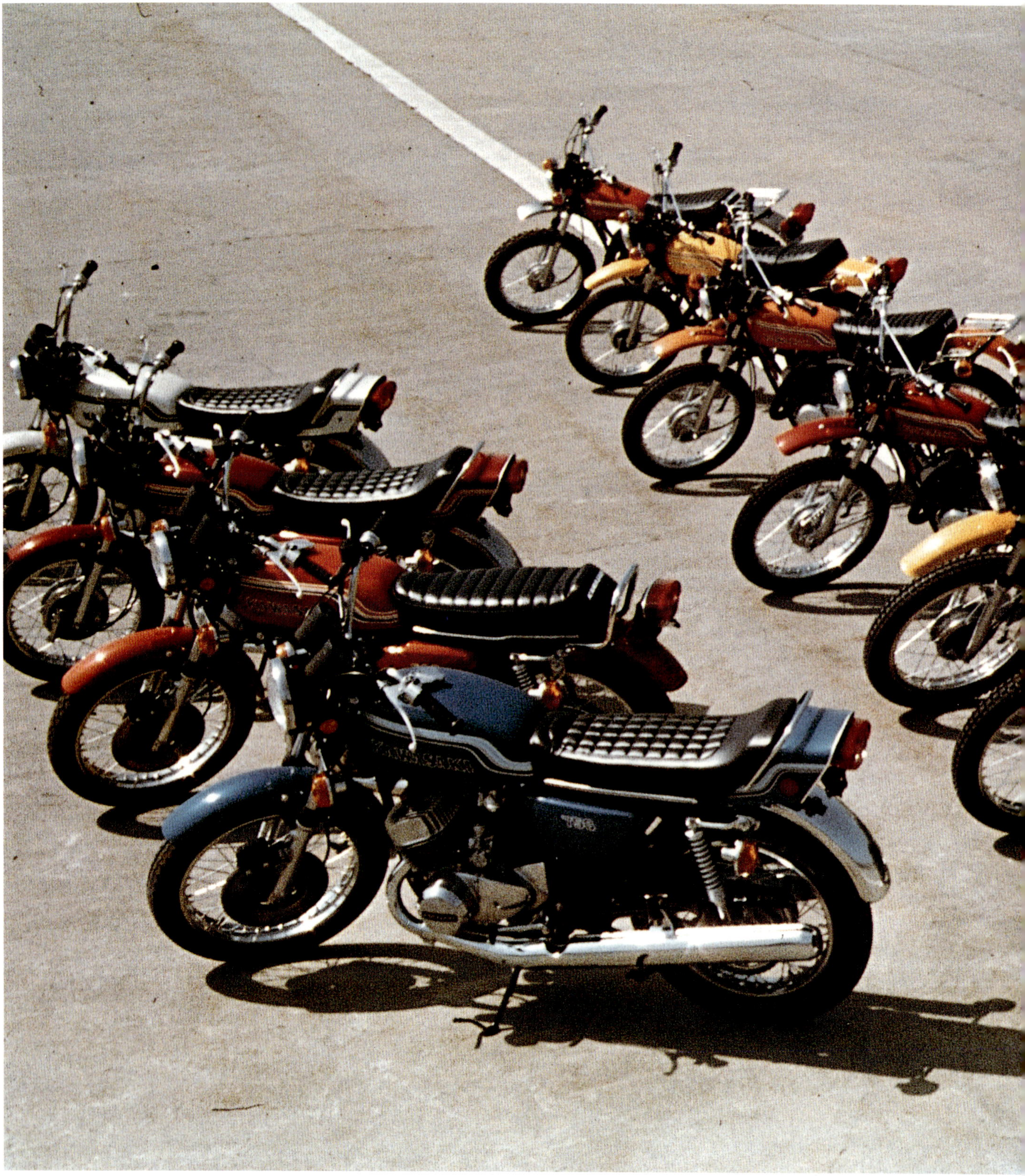

Above

A 1971 250 Samurai A1B disc valve twin. Finish of the fuel and oil tanks and nearside mounted tool box was in Pearl Ivory and white. The first Samurai model (A1) appeared in 1967; 1971 was the final production year. A larger model, the 350A7, was also offered

Left

The 1972 model line-up. Included were four versions of the three-cylinder (250, 350, 500 and 750), six trail bikes, two lightweight commuters; plus the MT1A monkey bike with its bright orange finish

company merged with Kawasaki Aircraft in 1963. The transformation of the corporate giant into a motorcycle manufacturer was complete.

Compared to Honda, Suzuki and Yamaha, who were already well established, both in the domestic market and abroad, Kawasaki came on to the scene very late. It is probably true to say that without its considerable financial and industrial strength, it is unlikely that the fledgling bike builder would have been able to rise to its current position of an equal member of the Japanese 'big four'.

The first real attempt by Kawasaki to enter the big-bike league was the 1966 W1. This was very much an inherited Meguro, albeit of 650 instead of 500 cc, itself a close copy of the BSA pre-unit A7/10 twin-cylinder models. The 360 degree parallel twin had pushrod operated valves and its

Left
A factory warehouse full of the fearsome 750H2A triples. This photograph was taken in 1973; colour options that year for the big 'stroker were either Candy Gold (shown) or Candy Purple

Right
A pair of 1977 KH400A4 triples outside a cafe situated just off the A1 trunk road south of Hatfield in Hertfordshire, England

74 × 72.6 mm bore and stroke dimensions gave a capacity of 624 cc. Kawasaki claimed 50 bhp at 6500 rpm. Other variants of the basic theme were the W1SS (1968-1971), W2SS Commander (1968-1970) and the street scrambler W2TT Commander (1969).

However, the motorcycles which really put Kawasaki on the map were a series of three-cylinder piston-port two-strokes. The first of these, the fearsome H1 (Mach III), appeared in time for the 1969 season. Its 498.8 cc (60 × 58.8 mm) engine produced 60 bhp at 7500rpm; this being progressively detuned over the following years until by 1976 it was down to only 52 bhp. But the original H1 was a flyer, capable of approaching 120 mph with ease; this was of course provided its rider had the nerve to hang on, as handling wasn't its strong point.

Along the way the H1 became the H1A (1971), H1B (1972), H1D (1973), H1E (1974), H1F (1975) and finally the KH500 in 1976.

If the H1 (Mach III) could be described as exciting then the H2 (Mach IV) was truly awesome. This followed the same basic two-stroke, three-cylinder engine layout, but now with a capacity of 748 cc (71 × 63 mm). Maximum power was 74 bhp at 6800 rpm. Performance, at least in a straight line, was the equal of any production motorcycle of its era. Unfortunately its 130 mph speed and searing acceleration were rather blunted by a low of twelve miles to the gallon when this performance was used to the full... The 750 H2 ran from 1971 through to 1975. Other

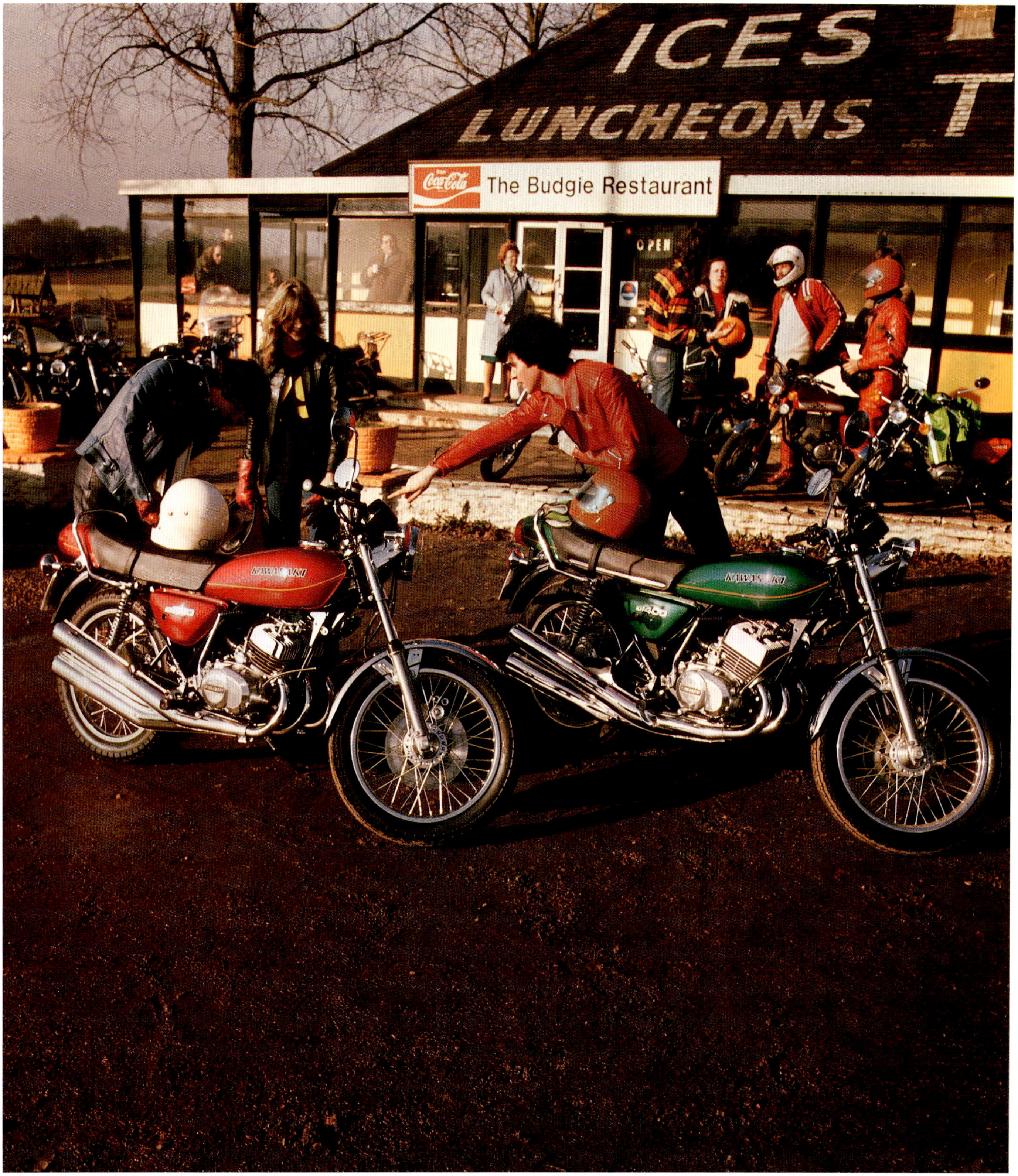
ICES
LUNCHEONS
Coca-Cola
The Budgie Restaurant
OPEN

Left

By 1978 the triples were coming to the end of the road; this is the KH250B3 in Lime green metallic

Right

In Britain, Kawasaki ran a successful race series for KH400 models. One of the leading contenders is shown at Snetterton in April 1979

Kawasaki triples included the S1 (249.5 cc) S2 (346 cc) and S3 (400 cc); the S1 became the KH250 in 1976, the S3 the KH400 in the same year.

The world fuel crisis of 1974 was what really killed the high performance triples, Kawasaki being forced to explore new avenues in the quest to establish itself as Japan's premier performance motorcycle builder; hence the transfer to multi-cylinder four-strokes as described in subsequent chapters.

A similar crisis – but from a different source – threatens the performance scene at the time of writing, as every motorcycle enthusiast knows. How will Kawasaki and the other manufacturers react to the threatened EC regulations limiting BHP?

RACE POSITIONS
3 1 23 6
10 10
37
46
KAWASAKI
SPEEDMAN
VICTORIA
ANDY

Lightweights

Like its Honda, Suzuki and Yamaha rivals, Kawasaki has offered a truly vast range of lightweight, commuter-type bikes over the years. In 'Big K's case, many of these have been powered by a rotary valve (disc valve) two-stroke engine – but not all some have had piston port, some reed valve, whilst others have even been four-strokes.

With such a wide range of models it is difficult to provide details of any more than the highlights. To my mind these were the KC90/100, Z200, AR50/80 and the AR125; the trail bikes are covered in another chapter. Others, such as the KH100/125, have also run for long periods but don't represent such milestones in the evolution of the lightweight Kawasaki as those I have chosen.

The forerunner of the KC90/100 series can be traced all the way back to 1965 and the 85 cc J1. Like the later bikes, this had a single-cylinder rotary valve two-stroke engine, with 4-speed gearbox. The first 90 cc model was the G1 of 1967, which became the GA1 in 1969 and the GA1A in 1971, followed by the GA2, GA2A, GA3, G2S and finally the G2T which ran from 1974 through to 1976. The KC90C1 first appeared in 1977 and was replaced at the end of the decade by the KC100. Besides its larger capacity the '100' had the advantage of a 5-speed gearbox. Power output was a respectable 10.5 bhp at 7500 rpm. Even ridden hard it would easily top 70 mpg, with over 100 mpg available if moderate throttle openings were used. Despite its camel-like thirst, *Motor Cycle News* achieved an electronically timed 66 mph in a 1982 road test.

The Z200 was a totally different type of lightweight. At its centre was a 198 cc (66 × 58 mm) four-stroke, single-cylinder engine with chain driven sohc. Its de luxe specification included a 206 mm (8.1 in) hydraulically operated front disc brake, electric starting (a back-up kickstarter was also provided), wet sump lubrication, 5-speed gearbox and a 12 volt battery/coil ignition system. The engine also incorporated a counter balancer to reduce vibes to a minimum. Later examples had new cosmetics, a revised silencer and transistorised electronic ignition system. Production ran from 1978 through to 1983, and it was very much a comfortable commuter/tourer

The original 1977 version of Kawasaki's long running KH125. Its 123 cc (55 × 51.8 mm) rotary valve two-stroke engine is air-cooled and has 5-speeds It was still in production in 1992

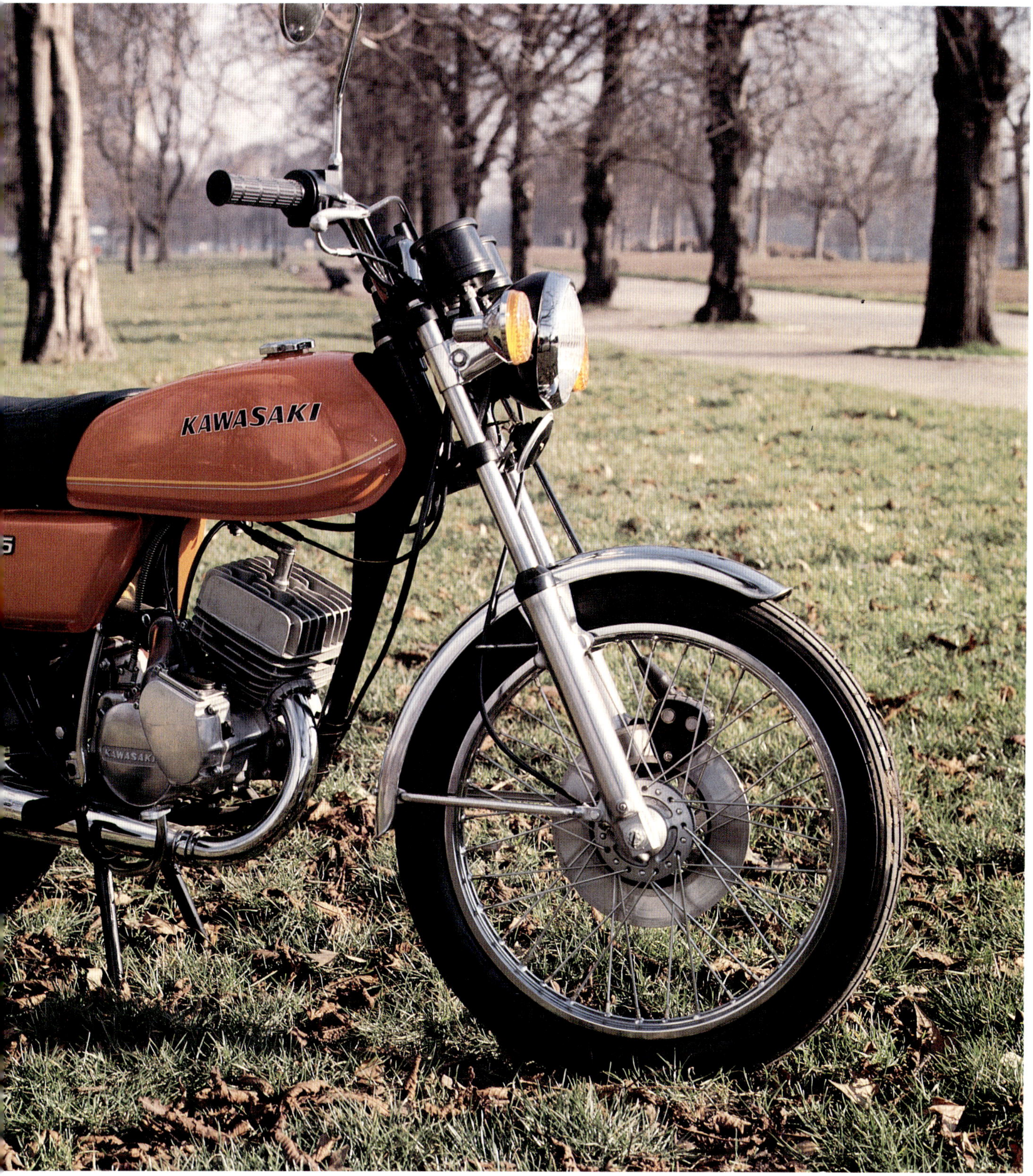
KAWASAKI

rather than a performance machine. There was also a larger version, the Z250C with cast alloy wheels, 246 cc and 19 bhp.

A complete contrast came in the early 1980s with the introduction of the AR50/80 series. Both these used a reed valve two-stroke engine. The AR50's measuring 49 cc (39 × 41.6 mm) while the larger AR80's 78 cc was achieved by a larger bore of 49 mm. leaving the stroke unchanged. Both models sported a 6-speed gearbox, although for certain markets (including Britain) a 5-speed gearbox was specified for the smaller bike, to comply with reduced power ratings. With the same 78 kg (172 lb) dry weight, the AR80 was by far the best performer, being capable of almost 125 cc class performance.

General specification for both the AR50 and 80 included leading axle front fork, Uni-Trak rear suspension, separate 'clip-on' handlebars, matching speedo/tacho, automatic (pump) oil injection, bikini fairing and 5-spoke cast alloy wheels.

Compared to such exotic 125s as Honda's Italian built NSR, Yamaha's TZR or Suzuki's RGV, the long running Kawasaki AR may seem a trifle ordinary. But not only is performance, at least on the restricted British market, not very different, but its purchase price (again in Britain) is far less.

The AR125 first appeared in time for the 1983 season, at its heart was a 123 cc (55 × 51.8 mm) liquid-cooled single-cylinder two-stroke which in unrestricted form gave 21 bhp (UK buyers got a choice of the full power version or a 'learner legal' 12 bhp).

The original 1983 model was coded A1, by 1992 this had become the A8, but remarkably in the nine years of production nothing much had changed except the colour schemes to mark the passing years.

A definite advantage of the AR125 for its mainstream learner riders is the Rotary & Reed Valve Induction System (RRIS). This broadens the power band and also improves fuel economy. Other features include: 6-speed gearbox, 18-inch cast alloy wheels, capacitor discharge ignition and side plus centre stands. Early models came with a bikini top fairing and belly pan, but now have a more comprehensive arrangement.

Over the years, Kawasaki lightweights have displayed the civilised side of motorcycling compared to their rivals – quite a contrast to its reputation for offering state-of-the-art performance machines further up the scale.

The KM100 proved popular in the 1970s. A 1979 A4 is shown; it was developed from the earlier 90 cc model

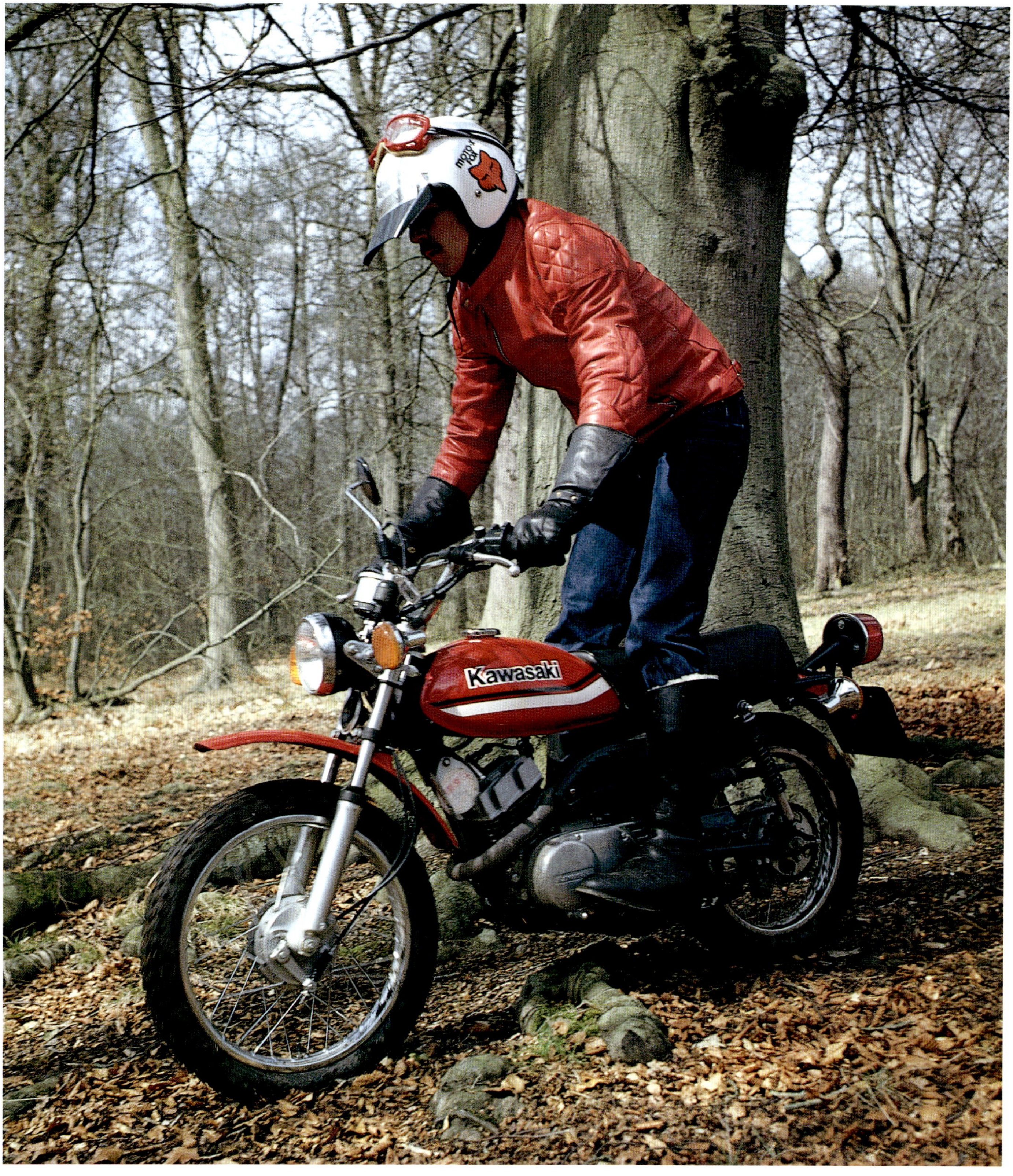
Kawasaki

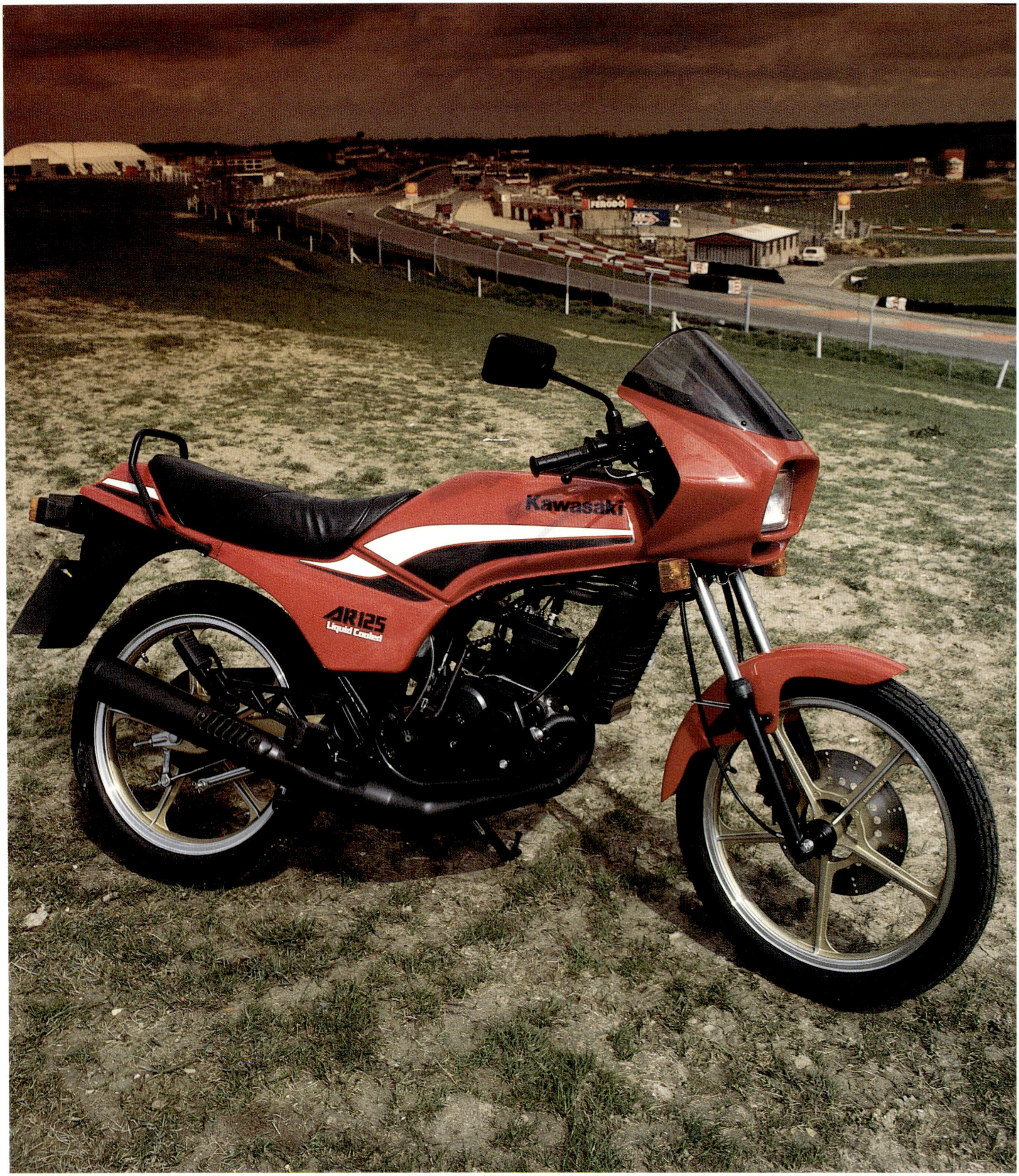
Kawasaki
AR125
Liquid Cooled
FERODO

Left
The AR125 (1983 A1 model shown) is a budget priced sports model with a wider spread of power than its rivals. This liquid-cooled two-stroke single features Kawasaki's Rotary & Reed Valve Induction System (RRIS). It has now been in production for a decade with little change

Above
AR125 engine. Bore and stroke are 55 × 51.8 mm respectively, giving a capacity of 123 cc; gearbox is a 6-speeder

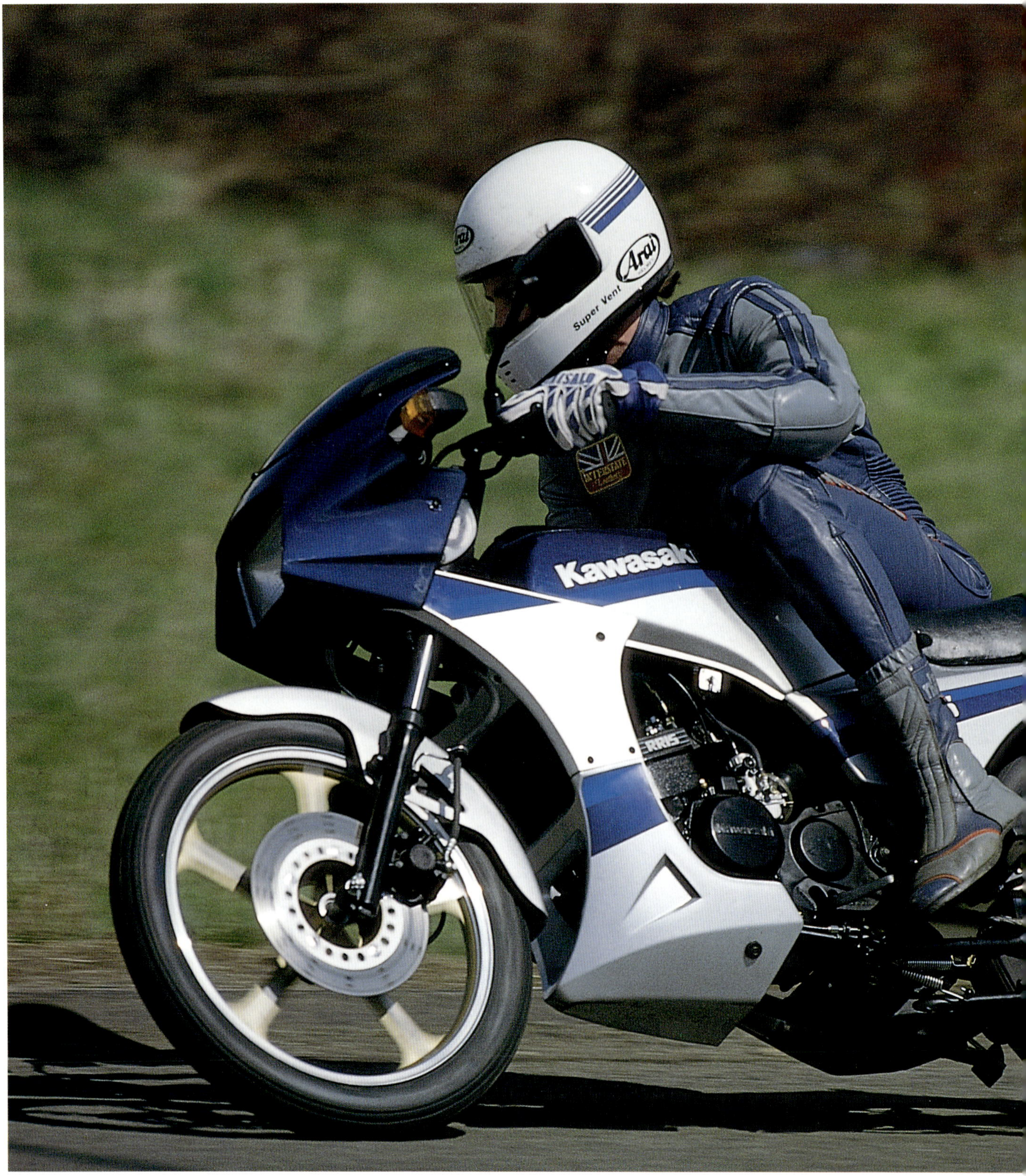
Arai
Super Vent
Kawasaki

Above
The AR80 is a large 78 cc (49 × 41.6 mm) version of the AR50. With the advantage of extra power and torque over the smaller engine, but being the same size, an excellent power-to-weight ratio provides performance not far short of a sports 125

Left
AR125B4 model of 1986 vintage. Revised graphics, colour scheme and fairing are the few changes from the original 1983 version

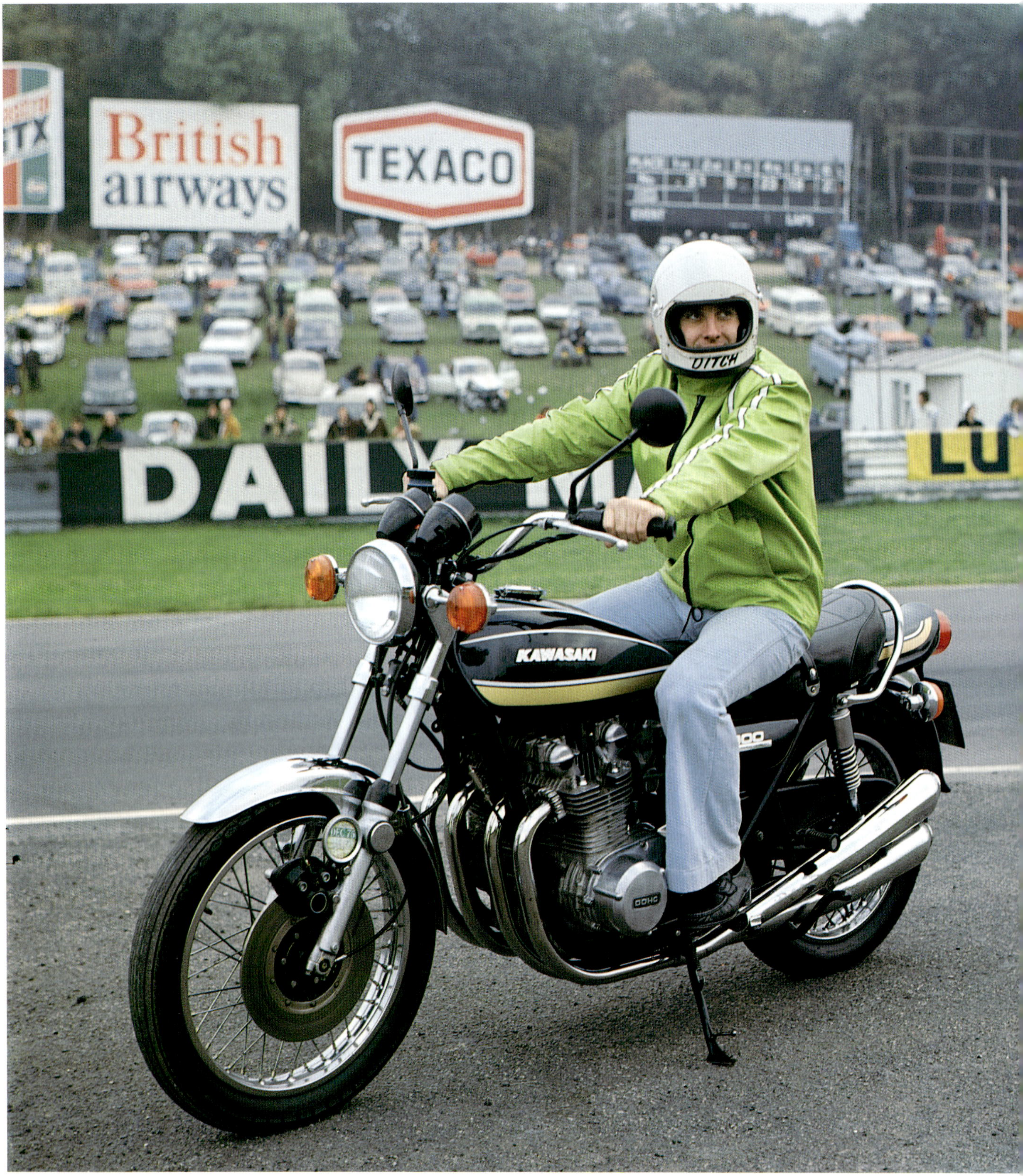
British
airways
TEXACO
KAWASAKI
DOHC

Air-Cooled Fours

If one motorcycle established the Japanese as builders of large capacity high performance muscle bikes it was the Kawasaki Z1.

Launched in 1973 the mighty 903 cc (66 × 66 mm) air-cooled dohc four-cylinder four-stroke re-wrote biking perameters. Okay, Honda's CB750 had been the world's first modern Superbike, but the Z1 was the king... absolutely the fastest and toughest streetbike in the world.

Work on the design which was to emerge as the Z1 had begun back in 1967 – before the legendary three-cylinder two-stroke Mach III had even entered production; the aim was to give Kawasaki a big-bore four-stroke replacement for the then current W series parallel twin.

Research had revealed a need for a sports tourer with a surplus of power, capable of hauling two people and full equipment over both short and long distances. It also needed to meet current and predicted future emission standards as the bulk of sales was aimed at the giant American market.

An early decision was taken that the capacity would be at least 750 cc and equipped with double overhead camshafts, necessary to provide the level of performance and sophistication four cylinders were required to achieve.

Any need to restrict specification was shattered by the appearance of Honda's CB750 four in September 1968 – and its resulting sales success only acted as a further spur to Kawasaki's design team.

The CB750 pioneered such innovations as the four-cylinder with sohc, 5-speed gearbox on a 750 cc class machine, and a disc front brake on a mass production model. Earlier attempts to market similar models had always been done on a very small scale making them hyper-expensive (the MV Agusta 600 for example).

By the end of 1969 the decision had been taken to use a capacity of around 900 cc and this with dohc, offered a substantial improvement in performance compared to the 750 cc sohc four-cylinder Honda.

A year later the first prototype was well advanced and in the spring of 1971 it ran at the Yatabe test track. But there were problems... the breather system let out more oil than it kept in, piston crowns collapsed, crankshafts failed. However, gradually, these and a myriad of other

Works racer Barry Ditchburn seated on a Z1-B at Brands Hatch in 1975

27

Left
The larger capacity Z1000 replaced the Z1 for 1977. An example is shown here at a drag race meeting in April 1978; standard except for after-market four-into-one exhaust

Right
Also new for the 1977 season: the Z650. This was Kawasaki's first 'small' four-cylinder machine; and it was so successful that it re-established this once popular capacity class

problems were ironed out. Along the way the design team learned a lot and was greatly encouraged with some truly excellent output and performance figures – 95 bhp and over 140 mph respectively.

With the USA being the major target, the next phase was a test programme in that country and early in 1972 two pre-production prototypes were extensively tested in the States. In June of that year several members of the world's motorcycle press were invited to Japan. They were shown a complete machine and their opinions and criticism sought. Production had commenced the previous month but only at very low levels. After the press visit it was full speed ahead and in September 1972 the Z1 was launched at the Cologne Show in Germany to a whirlwind of media attention and public amazement.

The machine was a clever blend of advanced engineering and practical features and, although it was blessed with double overhead cams, it was not over-complex and could be relatively easy to service. With the engine still in the frame the cylinder head and barrel assembly could be removed, as could components such as the alternator, starter, gear selector mechanism and clutch. Only if the crankshaft or gearbox required attention was it necessary to remove the engine from the frame – and this was a relatively simple, if muscle sapping task.

Although the rest of the Z1 was modern in appearance and specification, it was the engine which commanded the lion's share of attention. As if to confound the pundits it was not only capable of breathtaking performance, but also quiet, docile and able to run on regular grade fuel.

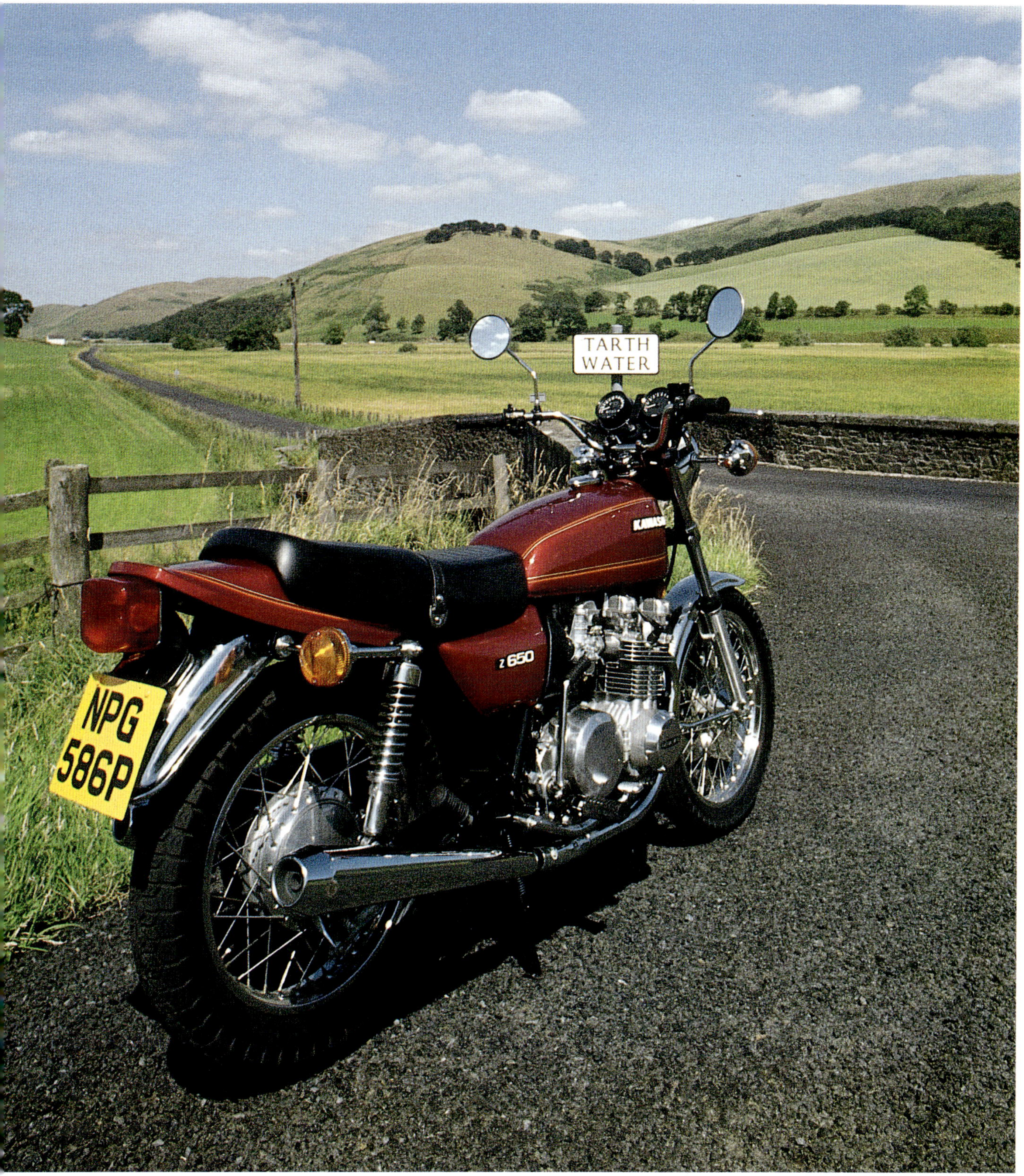
TARTH
WATER
z 650
NPG
586P

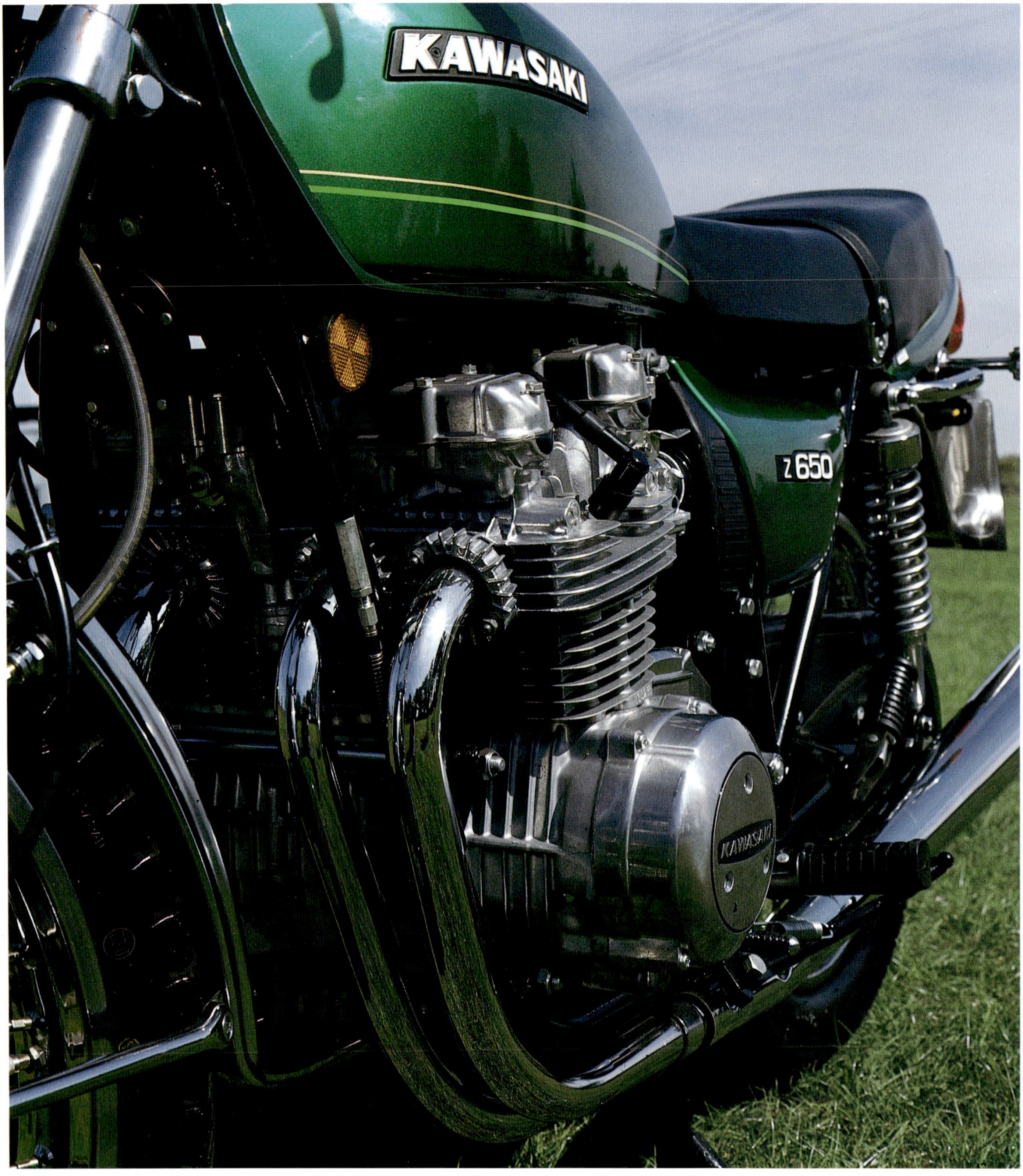
KAWASAKI
Z 650
KAWASAKI

Left
The dohc air-cooled four-cylinder powerplant of the 1977 Z650B1; actual capacity was 652 cc (62 × 54 mm)

Right
Over the subsequent years the Z650 was to prove one of the best loved and most reliable Japanese motorcycles of all time. A 1983 F4 model is shown here. Except for the addition of cast alloy wheels it was little changed from the original

The assembly was cast in the classic mould of the across-the-frame four already used on the factory racers from the likes of MV Agusta, Gilera, Benelli and of course Honda, with both their racers and CB750 roadster. The outer pistons moving together in opposition to the central pair. As was normal practice the crankcase split horizontally on the crankshaft and contained the gearbox.

The crankshaft was an all-caged roller affair with no less than six main bearings supporting it with the central pair clamped to the upper crankcase half. The crank was pressed together with the mainshafts forged with the webs and these shaped to provide the balance factor.

With a built-up crankshaft one-piece rods could be employed. The 3-ring pistons ran a compression ratio of 8.5:1; alloy cylinders carried pressed-in iron liners. The cylinder block featured a tunnel cast into its centre for the cam drive chain.

The valve seats were cast in the head and manufactured in an extra hard material enabling the engine to run on lead-free fuel. The valves themselves operated in bronze guides and this material proved to be one of the few errors that the design team made. Its wear rate was too fast for comfort and in later years it was substituted for iron, this becoming a retro-fit replacement for all Z1 motors.

Each valve used duplex coil springs. Valve adjustment was by the shim-set method. Both camshafts ran in four pairs of split plain bearings. The exhaust camshaft had an integral skew gear to drive the tacho.

The lubrication system was of the wet sump variety and took care of not

Above
The Z1-R was offered with the choice of two fuel tank size options

Right
The 1015 cc Z1-R was the performer in the 1978 Kawasaki range; maximum power was 90 bhp at 8000 rpm

just the engine, but the gearbox and primary drive. The main oil feed was connected to a pressure switch mounted in the upper section of the crankcase next to the engine breather. This was part of what Kawasaki termed Positive Crankcase Ventilation (PCV), which was already used in many car engines to reduce emissions.

Carburation was taken care of by four Mikuni VM28 instruments with round throttle slides and separate starter circuits. These were all mounted on a single plate enabling them to be removed as a complete assembly.

Ignition was by conventional coil and battery with two sets of points under the offside crankshaft end cover. The Z1 employed an electric starter (there was also a back-up kickstart) and a 12-volt system.

An interesting feature of the Z1 was the second oil pump; this was driven by the gearbox output shaft and supplied oil to the final drive chain. There was a small tank under the seat and lubricant reached its target via a non-return valve which passed it into the gearbox shaft. From there it was 'thrown' out through holes onto the chain as the shaft rotated.

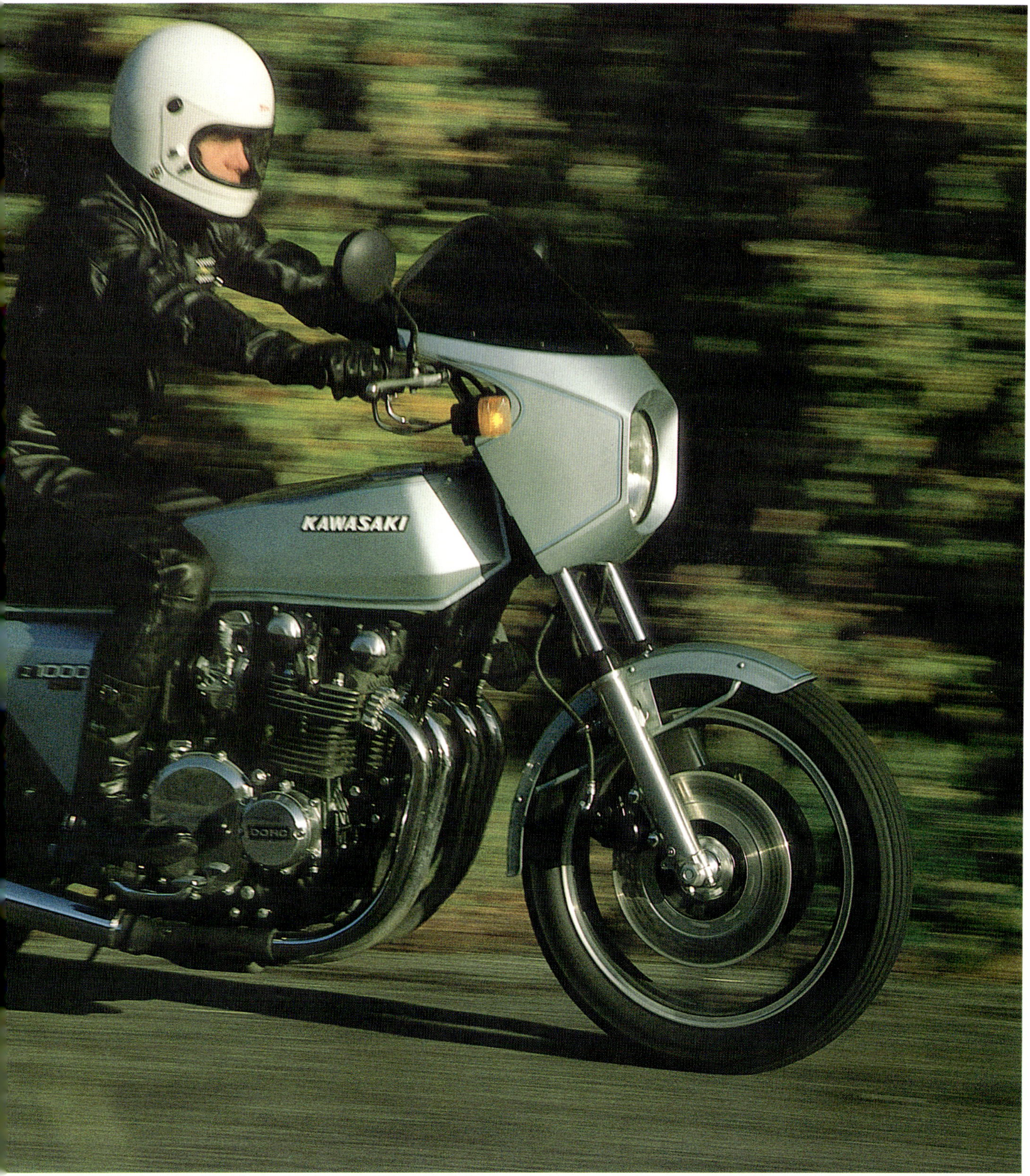
KAWASAKI

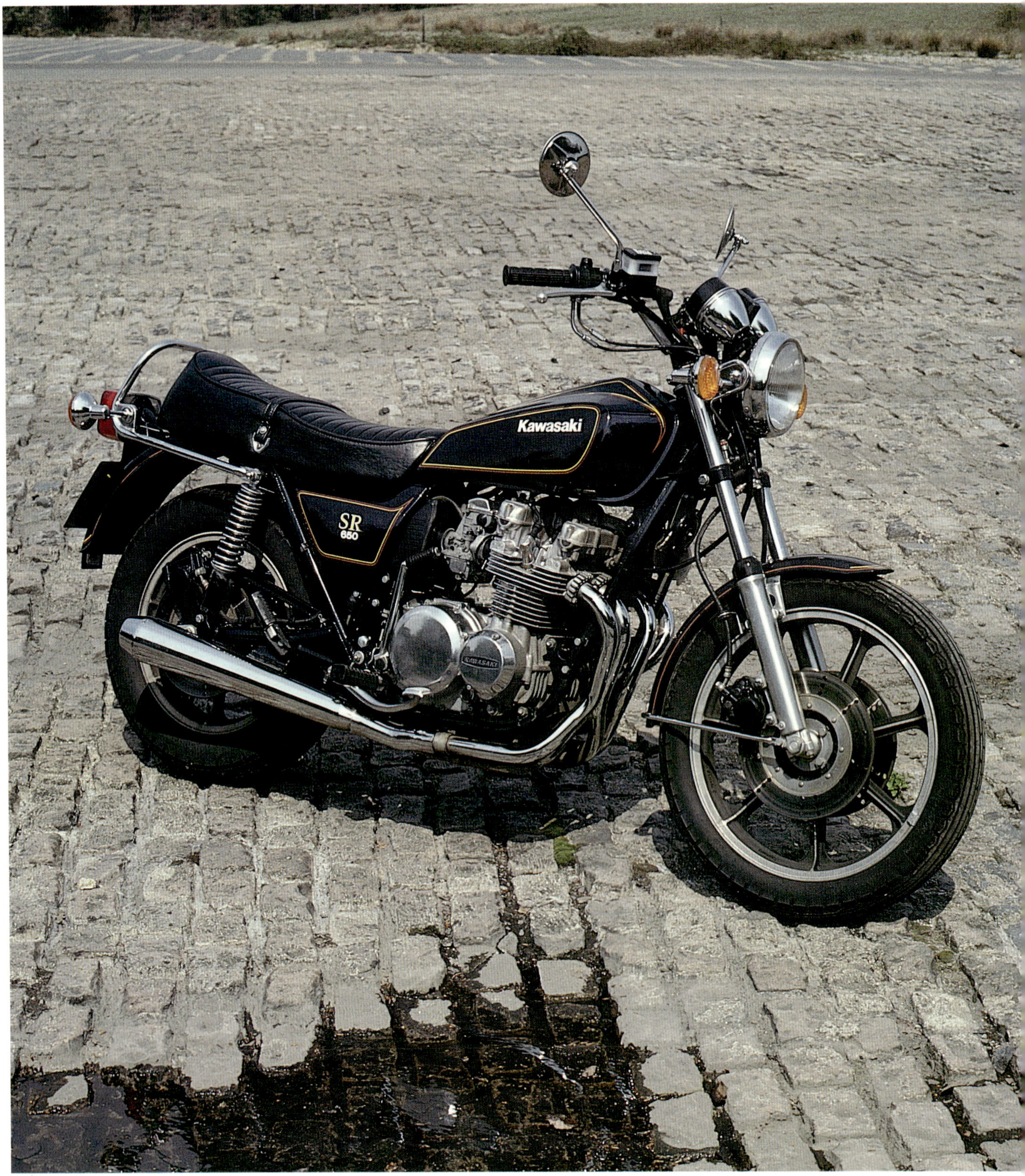
Kawasaki
SR
650
KAWASAKI

Left

A Z650SR of 1979 vintage

Right

A much-modified Z1000 competing in the 1979 French Bol d'Or 24-hour endurance race

Other details included exposed stanchion front fork, a 296 mm diameter hydraulically operated front disc brake and 200 mm single leading shoe drum rear brake, 19-inch front and 18-inch rear wheels, 18-litre fuel tank, direction indicators, centre and side stands and a dry weight of 230 kg (506 lb). Peak power was 82 bhp at 8500 rpm, giving a maximum speed of 130 mph. There was also a smaller 746 cc (64 × 58 mm) version for the Japanese home market. Final year of the 900 was 1976, by which time it was known as the Z900A4, built with dual front discs and audible turn signals.

During its four year production run, the Z1/Z900 had not only sold in large numbers all around the world, but had set new records for speed and acceleration wherever it went. For example at Daytona Week in March 1973 a trio of Z1s claimed 45 American and world speed and endurance records.

In 1977 Kawasaki's first 'small' four-cylinder machine appeared. This was the 652 cc (62 × 54 mm) Z650, and it was so successful that it re-established the capacity class. Over the subsequent years it was also to prove one of the best loved and most reliable Japanese motorcycles of all time.

In 1979, the four-cylinder Z500 proved that a mid-capacity machine with its advantages of slim lines and light weight could out perform much larger machinery. The Z500 displaced 497 cc with its 55 × 52.4 mm bore and stroke dimensions. A successful racing series for these machines was run in Britain during 1980 and was won by Neil Storey. A smaller version, the Z400, also appeared (399 cc – 52 × 47 mm bore and stroke).

The first four-cylinder 750 Kawasaki to sell on a world wide basis was the Z750E of 1980. Maximum power output from the 738 cc (66 × 54 mm) engine was 79 bhp at 9500 rpm.

Meanwhile the 900 had grown into a '1000'. This had been achieved by boring the cylinders out by 4 mm to 70 mm, giving a new capacity of 1015 cc. There was a wide range of models, including Z1-R, Z1000ST, Z1000H, Z1000J and Z1000R.

The Z1-R was very much in the cafe racer vogue with its aggressive lines and colour matched cockpit fairing – there was also a choice of fuel tank size. The Z1000ST was a luxury touring bike with a plush specification which included shaft final drive. Electronic fuel injection was a feature of the Z1000H; the J model was a budget priced Superbike and the R was a road-going version of the sit-up-and-beg racer campaigned by

Above
Black and gold paintwork was a feature of the 1980 Z1000H1

Right
Kawasaki's smallest air-cooled four: the 1980 Z400J1 at rest in an olde English setting; capacity 399 cc (52 × 47 mm)

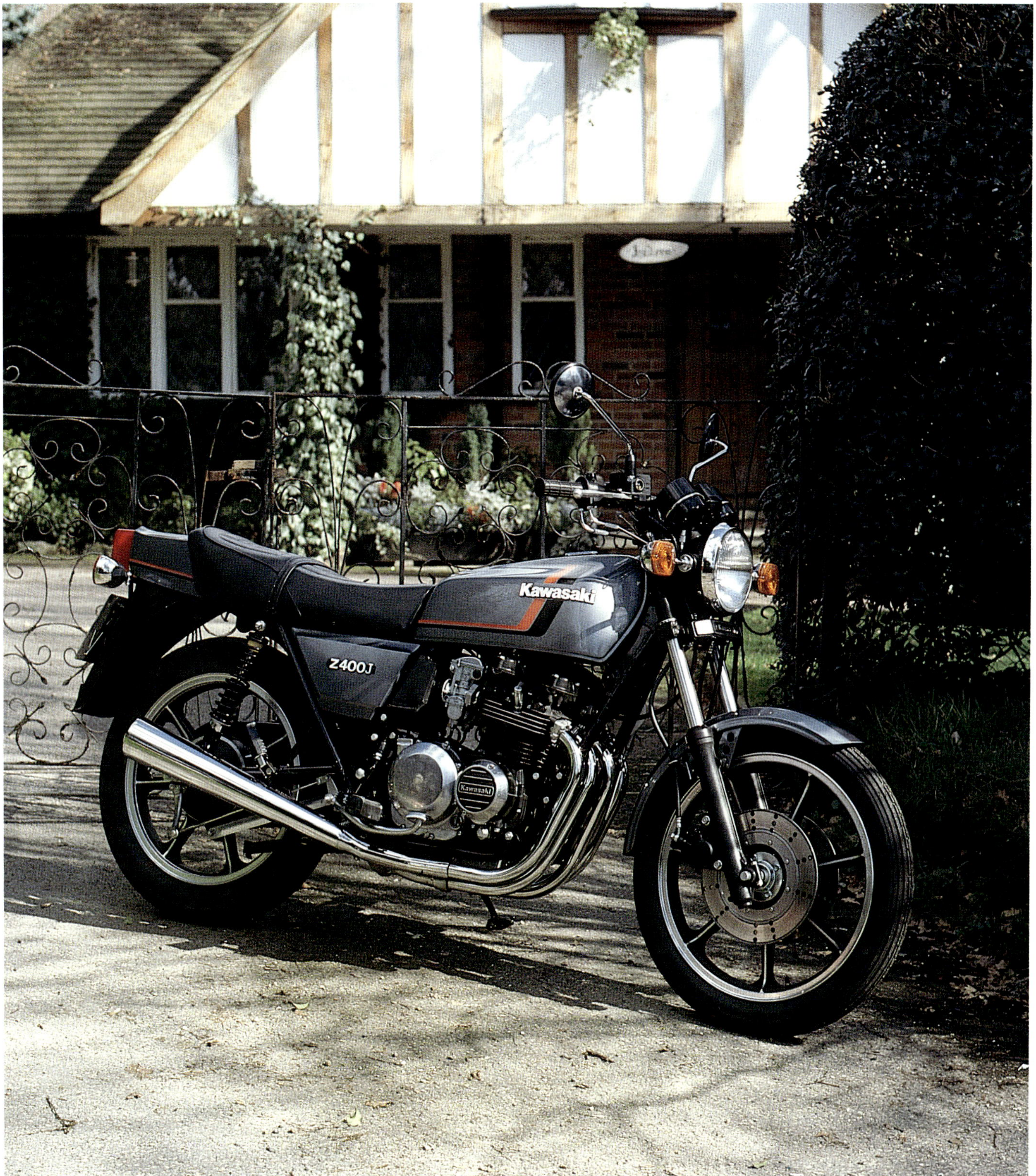
Kawasaki
Z400J

Eddie Lawson in American racing of the early 1980s, in the days before he became a grand prix star.

By 1982 the Kawasaki four-cylinder range had grown to include the Z550 (553cc) and Z1100 (1098cc), both these filling an important segment in the market. The most important one of the duo was, without doubt, the 550. The following year the company produced a master stroke: the GPZ range incorporating 550, 750 and 1100 versions. These sportsters – together with a 750 Turbo version which appeared shortly afterwards – were the pinnacle of Kawasaki's air-cooled fours, at least of the sporting variety.

Each represented class leading performance for its day: 550 – 61 bhp, 750 – 86 bhp, 1100 – 120 bhp, 750 Turbo – 112 bhp.

Testing the GPZ750 Turbo for *Motorcycle Enthusiast* in June 1984 the author said: 'Right from the first moment I sat astride the Turbo and

A successful racing series was run for the new Z500 in Britain during 1980. One of the leading contenders was Neil Storey, seen here at Aintree on 19 July of that year

Late in 1980 the Z500 became the Z550 by simply increasing the bore size from 55 to 58 mm, giving a new capacity of 553 cc

pressed the starter button I felt comfortable and part of the bike, a good feeling, which was heightened when moving off. The dog leg levers made control easy and in coping with the Oxford traffic I found the motor both flexible and docile at low rpm; both feet could be put flat down on the road when at a standstill. But any original doubts that I might have harboured were soon dispelled out on the open road. I soon realised the true advantages of a well designed turbo; tremendous performance, coupled with a much superior power to weight ratio, a mixture of docility and fire dependent on whether you are on or off the turbo'.

The 1979 Milan Show had seen the debut of a very different breed of Kawasaki four, the GT – in 750 guise. This was followed in early 1983 by a smaller 550 version. Both shared the title of what *Motorcycle Sport* labelled 'A good, practical motorcycle'. And then went on to say 'Whoever thought the day would come when a powerful Japanese Superbike would

Above
Forerunner of the highly successful GPZ550, the GP550 of 1982 with 60 bhp and Uni-Trak rear suspension

Right
A 1983 Z550F. This machine offered a cheaper purchase price than the GP/GPZ models, but retained much of the performance and handling of the more glamorous models

earn the epithet "traditional"? The Kawasaki GT750 represents a salutary lesson in what has either gone wrong (or right) in modern motorcycle design. Yes, its looks are a bit dull compared to the shining plastic state-of-the-art machines that top each manufacturers range. The fuel tank holding a whopping 5.3 gallons (24.3 litres) is in the place where fuel tanks are supposed to be. The riding position is "touring" rather than "racing", the seat is comfortable and can take two normal people, and the suspension has units either side. Why, you can even see the engine!'

Perhaps this is why both GTs have survived for so long virtually unchanged: there is still a steady market for traditional motorcycles – and the GT series had the added advantage of shaft final drive.

This brings us almost up to date, and the Zephyr 'retro' models. Born very much from the success of a specially produced American-only 550 model, there are now three versions; the original, plus the 750 launched in 1991, and the 1100 which came out a year later.

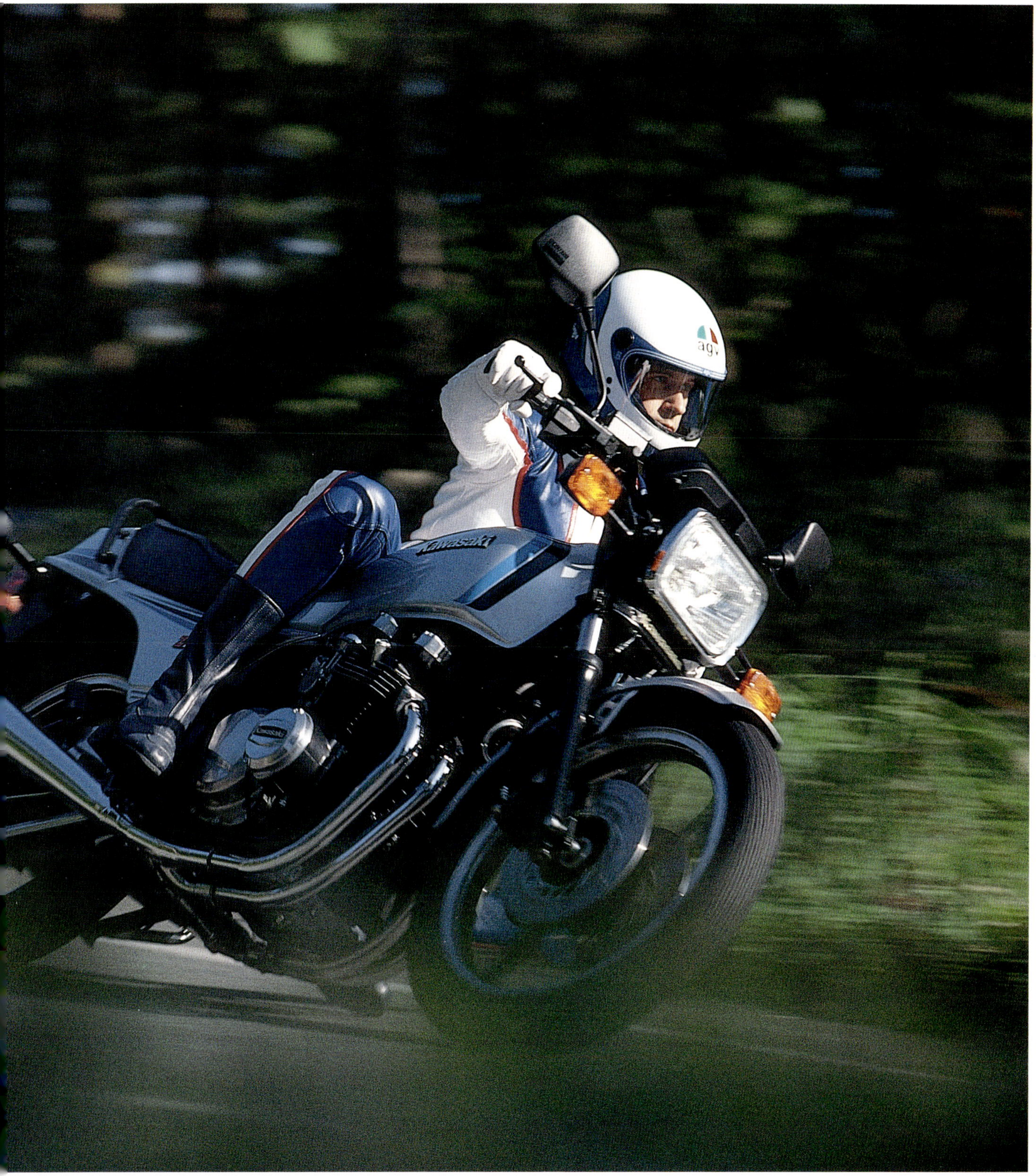
agv
Kawasaki

Above
Launched at the end of 1981, the 750GT has been a consistent best-seller ever since; its conservative styling and shaft final drive have proved popular down through the years. Once again, Kawasaki found the right formula first time

Right
Bend swinging on a 1982 Z750L2. This was the base model with chain final drive and no fairings

Don't make the mistake of imagining that because they look like a throwback to the past that technical development has not been taken into account, or that somehow the only people that ride them are those living in a time warp.

Compared to the Z1100R of 1984, the Zephyr 1100 has a wider bore and stroke (73.5 × 62.6 mm, compared with 72.5 × 66 mm) and retains two valves per cylinder (albeit with dual valve springs), using the same shim-set adjustment as the earlier bike. Both are centre cam chain designs, but the Zephyr has two spark plugs per cylinder. Why? Kawasaki engineers found that with such a wide bore and large valves, a single plug would not give quick, complete combustion of the fuel charge; thus dual spark plugs were found to be necessary, in much the same way that the Italian car company Alfa Romeo discovered, and consequently introduced their much-vaunted 'Twin Spark' design.

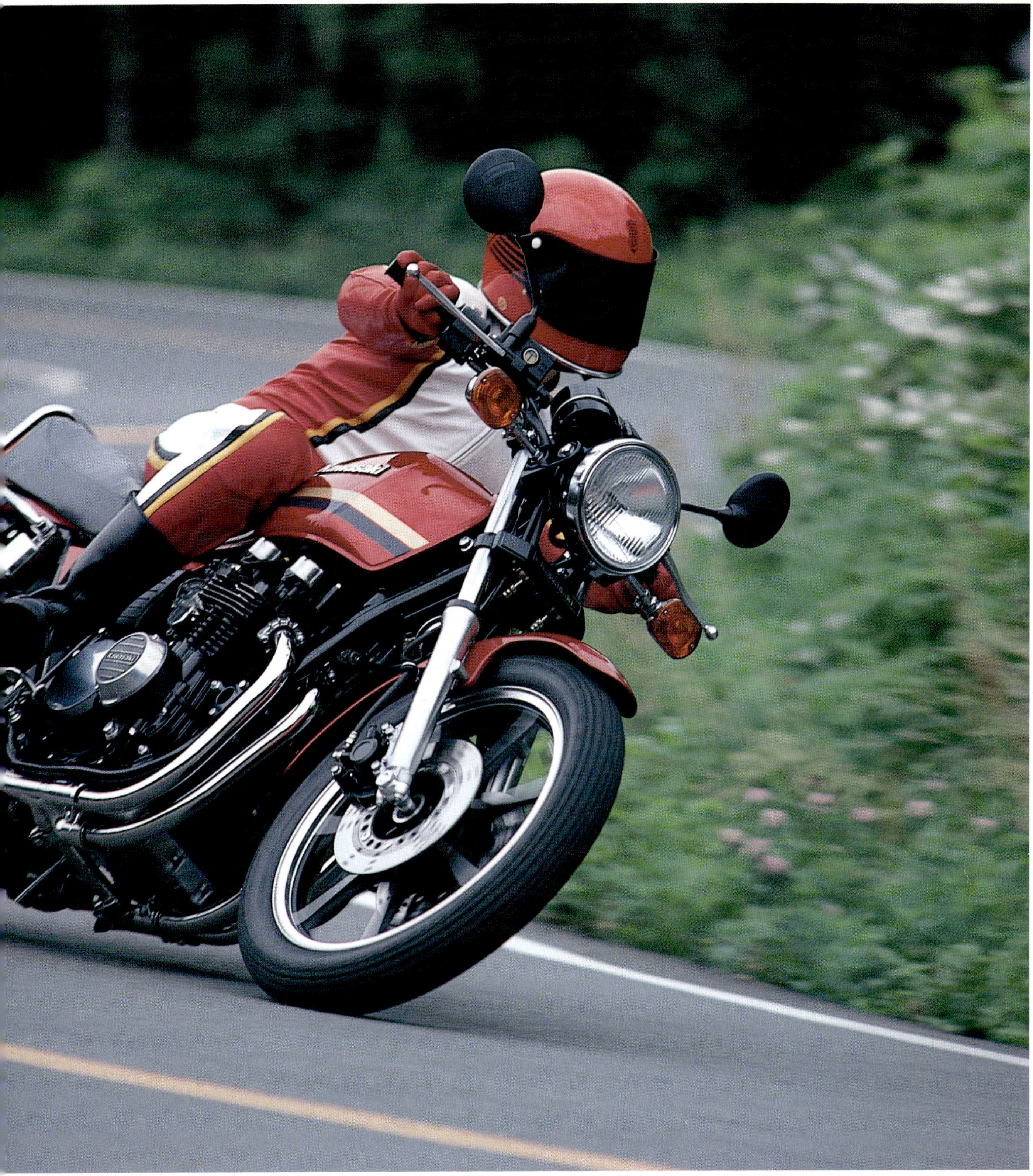

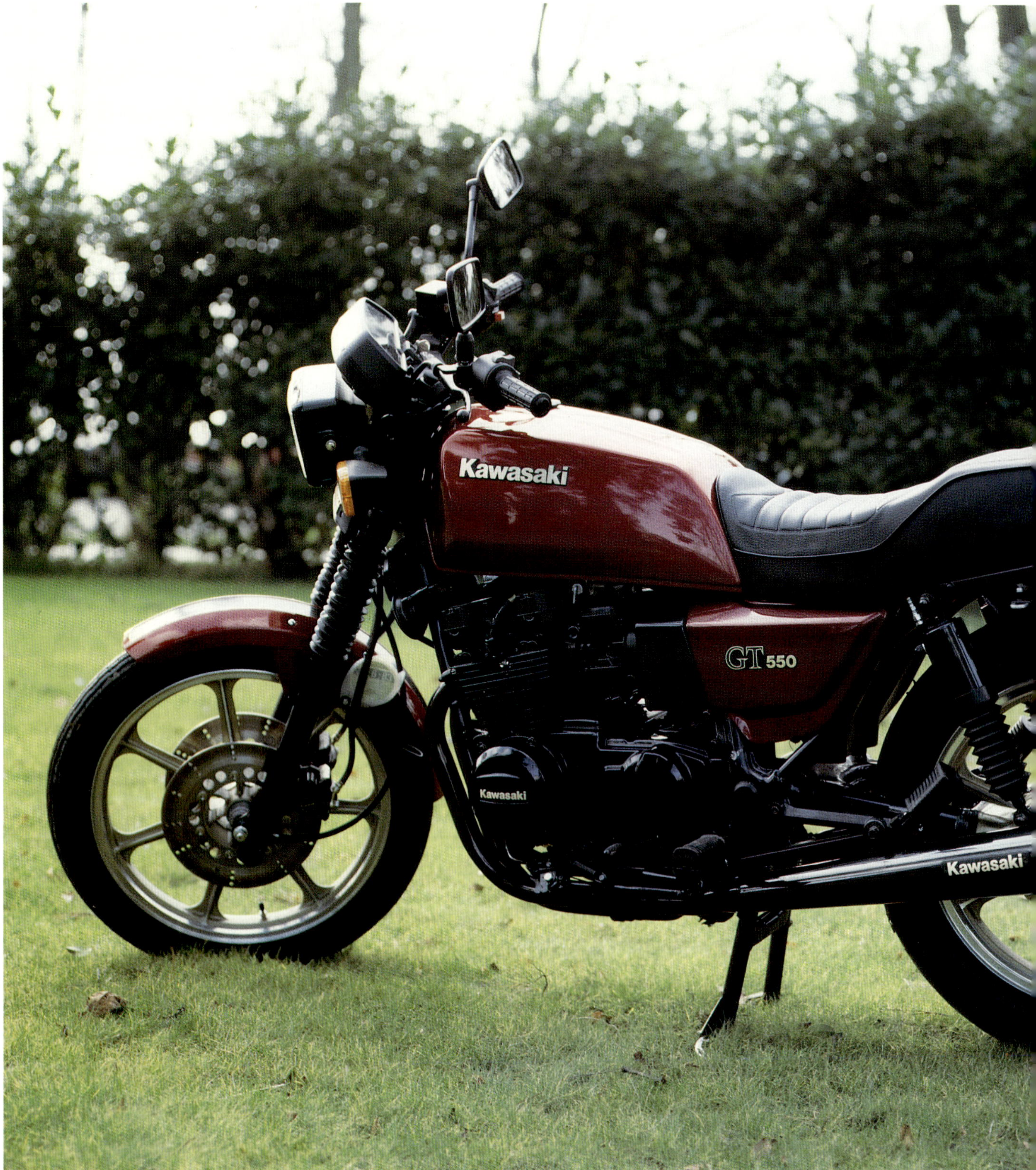

Kawasaki
GT550
Kawasaki
Kawasaki

Above
'Retro' 750 Zephyr offers potential owners a glimpse of the past, but with the technical advantages of today; there are also 550 and 1100 versions

Left
The 750GT was followed by the smaller 550GT in 1983. This is the original G1 model; the latest G7 is remarkably similar

Kawasaki

GPz 1100

Model: ZX1100 A1 Engine: Four cylinder four stroke of 1089cc capacity, DOHC. Max power 120 hp @ 8,750 rpm. Five speed gearbox. Suspension: Front: air adjustable tele forks with anti dive. Rear: air adjustable Uni-Trak. Wheels: Front 18 in, Rear 17 in. Brakes: Triple disc. Dimensions: Wheelbase 1565mm, seat height 800mm, weight 244 kg, fuel capacity 20.4 litres. Colour: Sunbeam Red.

WHO CAN CATCH A KAWASAKI?

Left
The GPZ1100...

Above
... and the GPZ550; both were their respective class leaders in the sales charts for the 1984 'season'

Trial bikes

The first modern dual purpose motorcycles were the 'Street Scramblers' of the 1960s; in reality being little more than lightly modified roadsters. By the end of the decade a new breed of motorcycle was about to be born; the Trail Bike, the craze came about thanks to an expanding Stateside off-road market.

Americans had always gone for cubes and this was reflected in their preference for either Harley-Davidson V-twins or large capacity British vertical twins. Then they discovered off-road riding and with vast tracks of open country at their disposal it was soon the fun thing to do. They also quickly realised that heavyweight roadsters were not the ideal tool for such excursions... This in turn led to a surge of interest in lightweight single-cylinder two-strokes which were fast enough to keep the rider happy, but easy to handle over mixed terrain. It was also simple to hitch such a bike on the rear of a camper and go off for the weekend with a full range of creature comforts.

Both Harley and the British were slow to cater for this new trend; the Japanese, including Kawasaki, were not. 'Big Ks' only mistake came by originally allowing the American Eagle company to market its range of trail bikes. However, very soon they established their own set-up and never looked back. The year 1969 witnessed a plethora of new models including a pair of disc valve 250s, a pair of 100 cc ultra-lightweights and an exciting newcomer to the scene, a full size 350.

Coded F5 and known in the States as the 'Bighorn', the newcomer was capable of a fine performance thanks to its 346 cc (80.5 × 68 mm) 33 bhp disc valve motor and 5-speed gearbox. The Bighorn boasted a full duplex frame and Hatta front forks. These latter components were named after the Kawasaki engineer who designed them. Both wheels featured alloy rims with the front of 21-inch diameter – the first of this size on a Kawasaki trail bike.

A 350 trail bike was listed until the 1975 season, but from 1972 onwards it was the F9, not F5. The F9 was also less powerful (28 bhp) than its predecessor, this following a trend set by the H1 (Mach III) sports roadster.

Other early Kawasaki trailsters included the 175 F3 (Bushwhacker), 175 F7, 250 F4 (Sidewinder), 250 F8 (Bison) and 250 F11. Although the 350 didn't continue the 175 became the KE175 (1976) and the 250 the KE250 (1977).

From 1971 Kawasaki had added a 125 trail model, the F6. This in turn

Above right
One of Kawasaki's earliest trail bikes was the F5 Bighorn, a 346 cc disc valve two-stroke single which was built mainly with the American market in mind; it was offered between 1969 and 1971

Right
A feature of the early Kawasaki trail bikes was their excellent disc valve two-stroke engines, typified by the KE175 (1977 B2)...

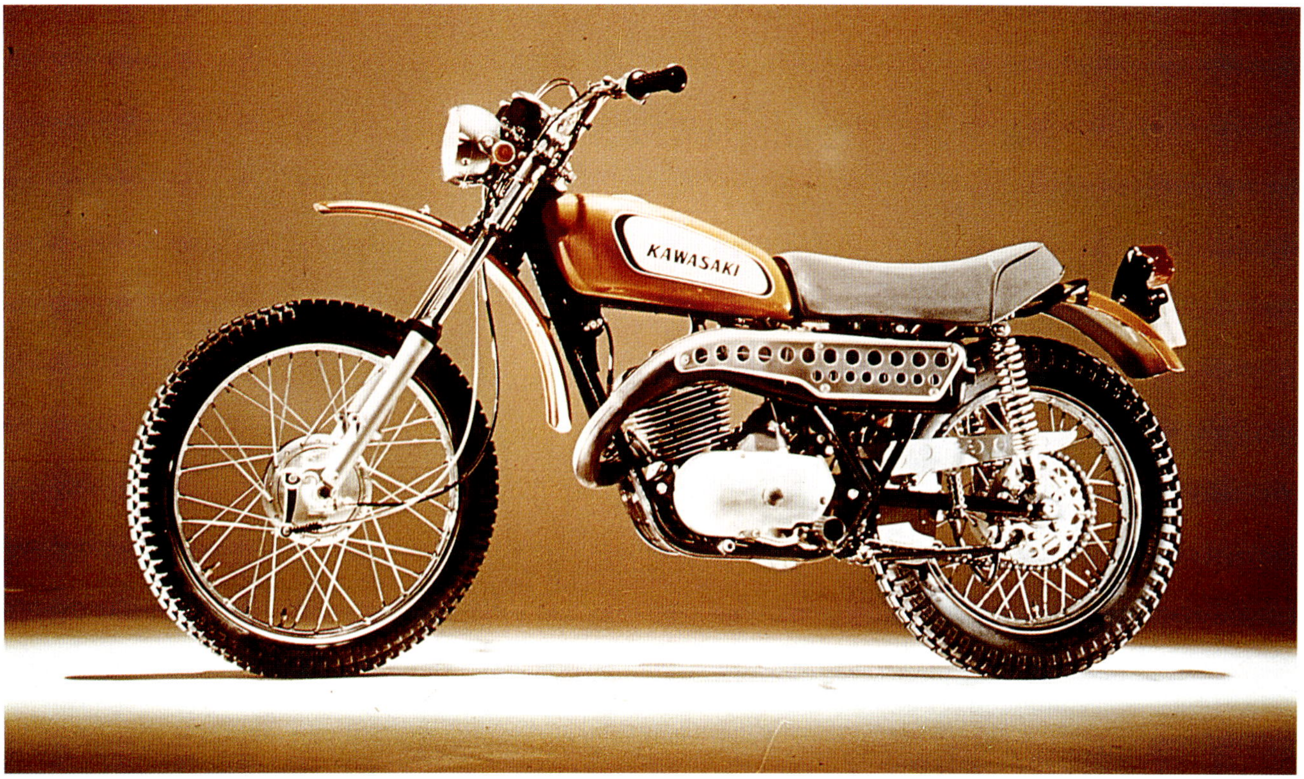
KAWASAKI

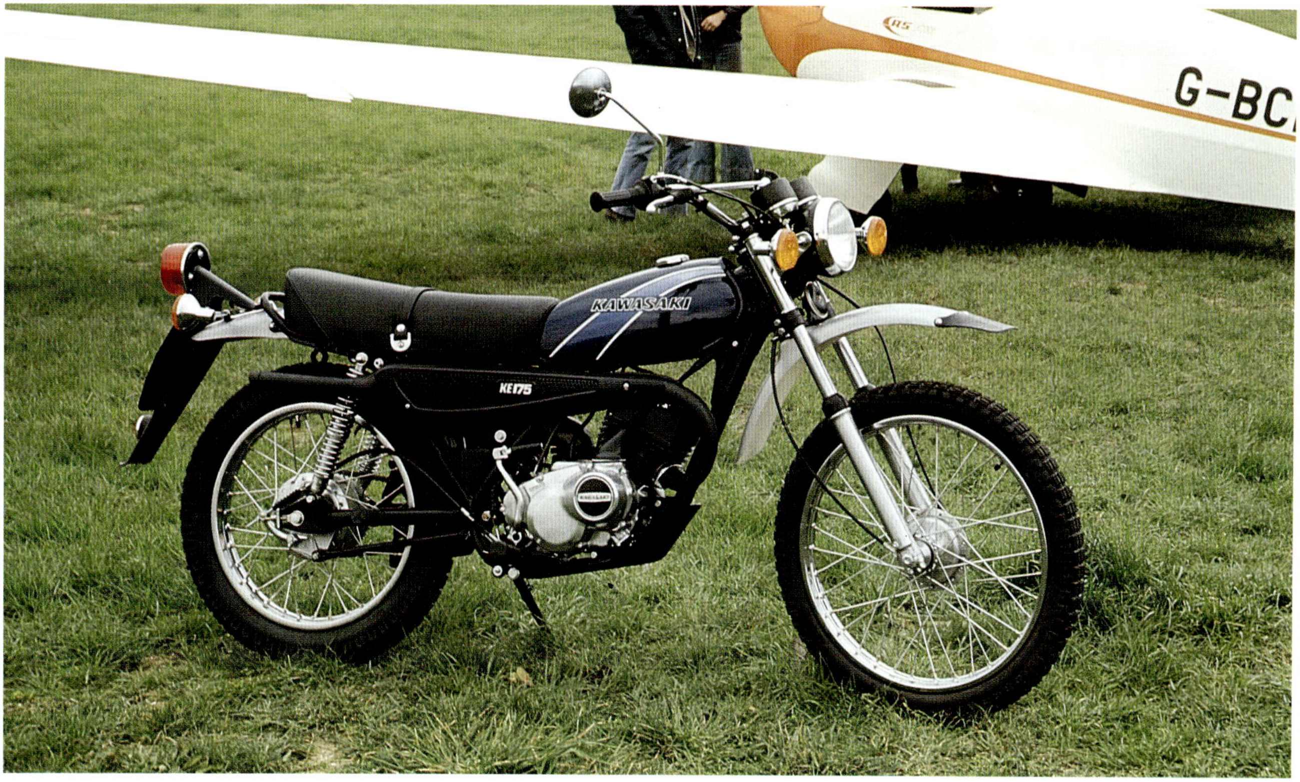
KAWASAKI
KE175

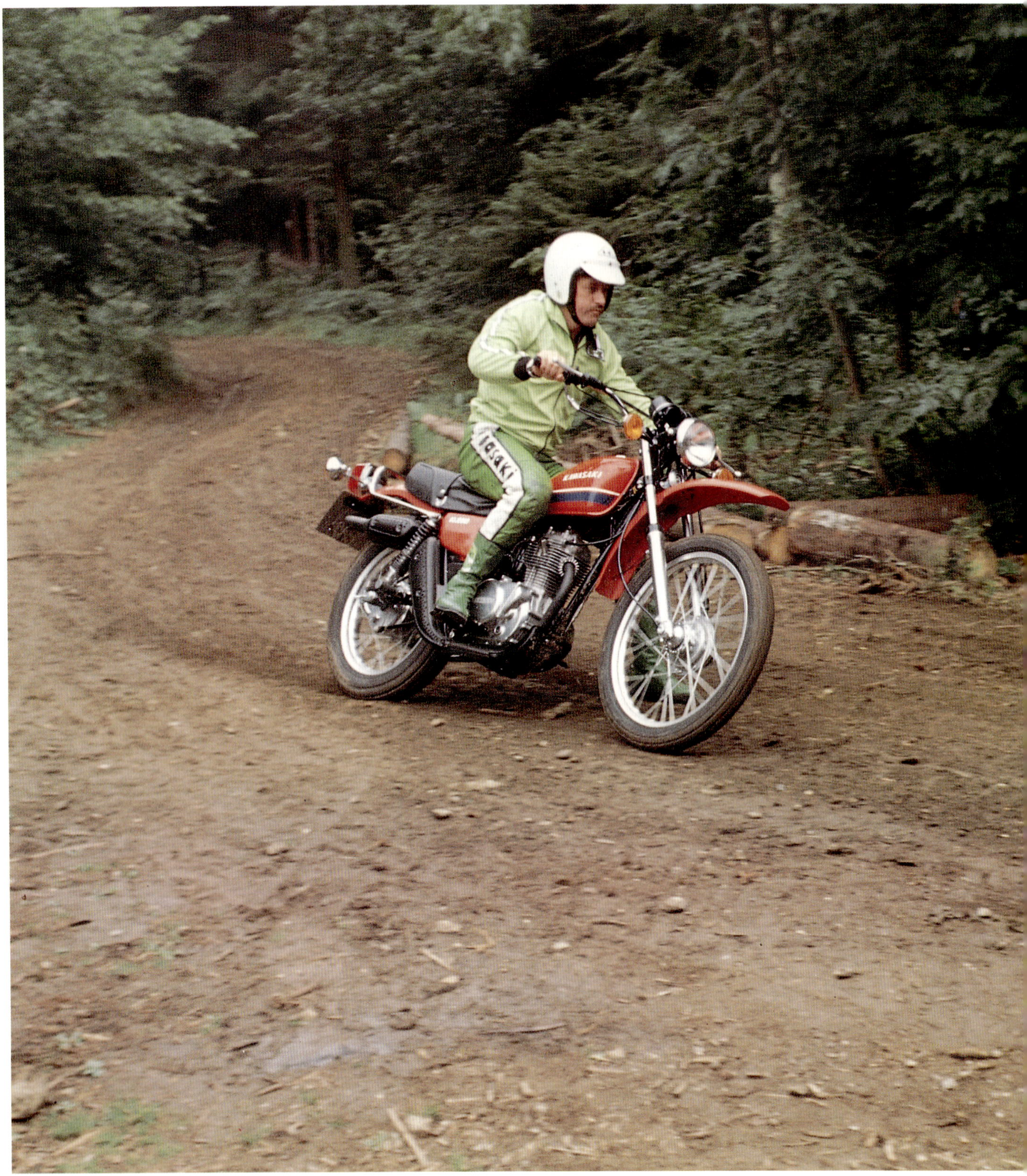

Above

... and the smaller KE125 (1978 A5)

Left

First of the many four-stroke trail bikes from the Big K stable was the 1978 KL250A1. This featured a 246 cc (70 × 64 mm) single-cylinder, sohc engine and 5-speed gearbox

became the KS125 in 1976 before finally becoming the KE125 in 1976.

A feature of all these early Kawasaki trail bikes was their single cylinder disc valve two-stroke engines.

The KE line ran for a number of years proving not only popular in the showroom, but reliable performers both on the street and the dirt.

Next came the first of many Kawasaki four-stroke trail bikes, the 1978 KL250A. This had a 246 cc (70 × 64 mm) single-cylinder, single overhead cam engine and 5-speed gearbox. Maximum power was 21 bhp at 8000 rpm. Other technical details included primary geared kickstarter, CDI ignition, auto cam chain adjuster and fast idle carburettor.

The KE and KL were to survive virtually unchanged into the 1980s. The 175 and 250 KEs were the first to go, replaced by the likes of the 124 cc (54 × 54.4 mm) liquid-cooled motocross style hi-tech KMX125 (1986) and eventually the KDX125 (1990). Production of the 99 cc (49.5 × 51.8 mm) KE100 continues today, while the final version of the

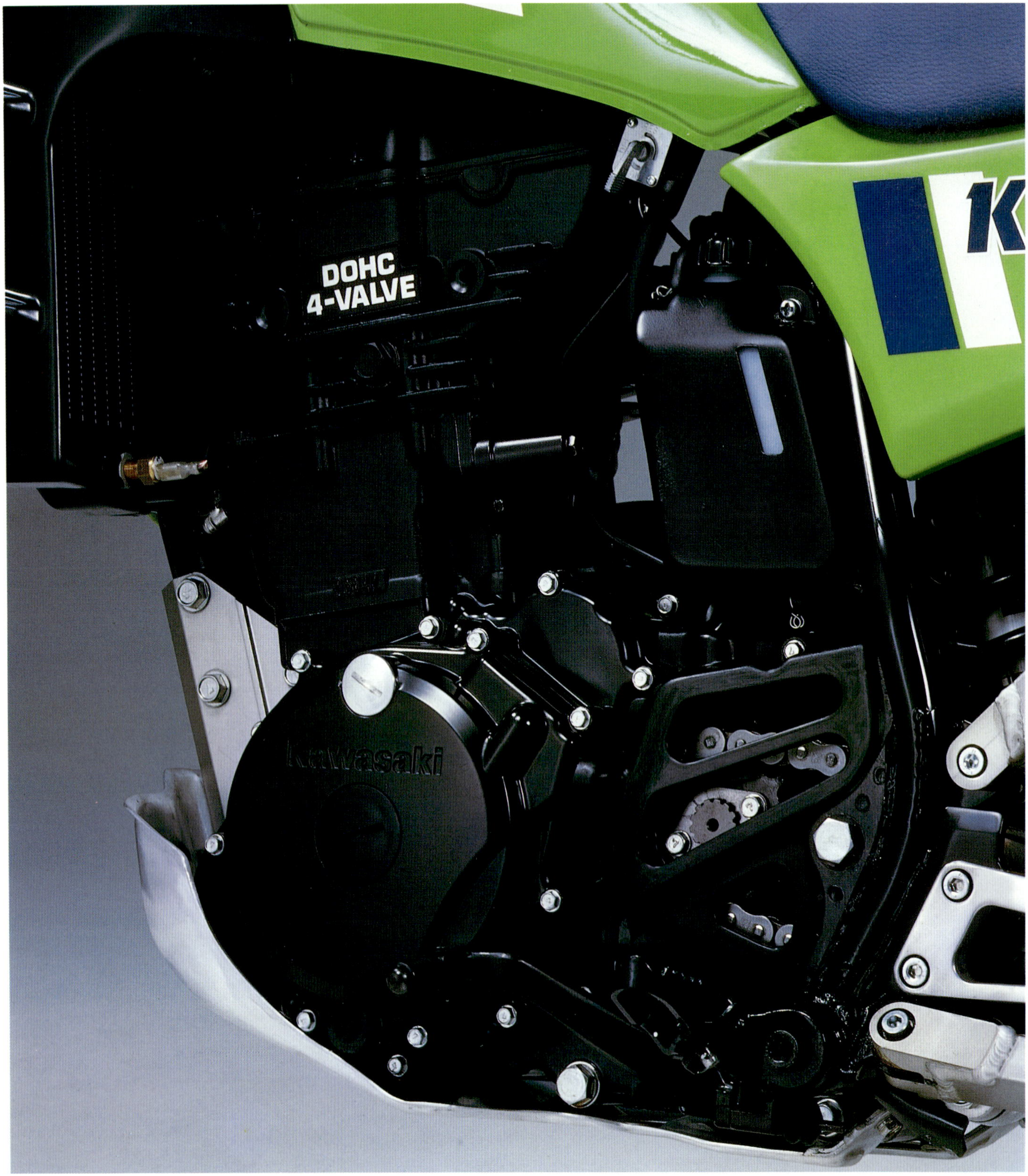
DOHC
4-VALVE
Kawasaki

Liquid-cooling was a feature of the dohc KLR600 of 1984

long running KE125 was eventually axed in the late-1980s. There was also a 191 cc (67 × 54.4 mm) KMX200.

On the four-stroke front, the KL250 was replaced by the 249 cc (74 × 58mm) double overhead cam, 6-speed KLR 250 in 1984. The KLR theme was continued with first the 546 cc (96 × 78 mm) KL600 in 1984 and later by the 651 cc (100 × 83 mm) KL650 in 1987.

Another feature of the KLR series is their 4-valve cylinder heads; the 600 and 650 having liquid-cooling and 5 instead of 6 gears like the smaller KLR250.

1989 saw the arrival of the Tengai, which translated means 'edges of the world', and with Italian styling. In may ways it aped Honda's Transalp, but with a single-cylinder rather than V-twin powerplant.

Engine-wise the Tengai was the same as the KLR650; both being equipped with a built-in automatic compression release and electric starter as standard.

A large 23-litre fuel tank, comfortable dual seat, together with the frame-mounted fairing with tinted screen plus handguard allows the rider to travel long distances between stops. There are even built-in rear carrier and frame-mounted pillion footrests.

For 1991 Kawasaki brought out its first twin-cylinder trail bike. This was the KLE500 which employs the same basic 498 cc (74 × 58 mm) dohc engine found in the GPZ500S and EN500 models. For use in the KLE, Kawasaki revised both the intake and exhaust systems to deliver more mid-range power.

Massive 41 mm fork stanchions provide 220 mm of wheel travel and the single shock Uni-Trak rear suspension features 5-way adjustable spring preload.

But the most striking aspect of the KLE500 is its futuristic quarter-fairing and slant 'bug-eye' headlamp; a bonus is that it provides a surprising level of rider protection. Other features include a neat instrument console with well lit instruments, rear carrier, hand protectors, fork gaiters, disc guards, sump guard and a range of striking colour schemes. It must be said, however, that the KLE is very much a roadster rather than a dirt iron. In fact you go off-road at your peril!

Above
Kawabunga! 600 dohc Kawasaki trail iron was good on both the street or dirt, providing an excellent combination of handling and gutsy four-stroke power

Right
Motocross styling is evident in this 1986 KMX125 which is being put through its paces off-road

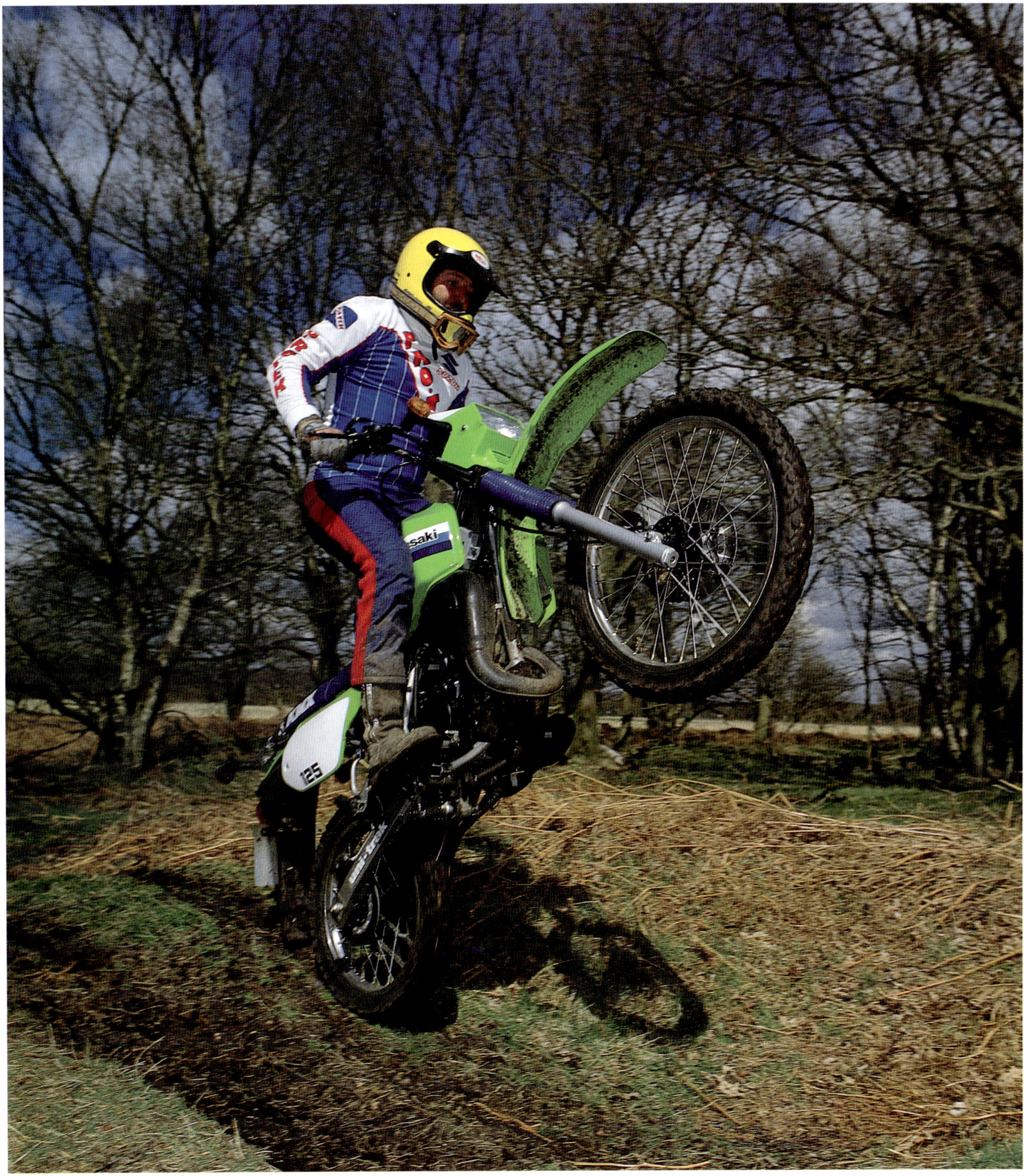
125

Racing

As revealed in the earlier chapters of this book, Kawasaki was a latecomer onto the motorcycle scene.

Kawasaki Heavy Industries took over the old established Meguro marque in 1963. The latter company had campaigned a 500 cc ohc single with some success at the annual Asama dirt track races in the late 1950s, while the Kawasaki emblem was first seen on the tank of a 125 cc machine which carried off the 1963 Japanese motocross title. However, the corporate giant realised that to make any real impact, it would have to take a leaf out of the Honda, Suzuki and Yamaha book: go for broke and aim for a world road racing championship.

The first ever Kawasaki to appear on the grand prix circuit did so in the 125 cc event at the Japanese round at Suzuka in October 1965. In the event, all three of the disc-valve, watercooled, two-stroke twins retired.

For 1966, Kawasaki not only improved their 125 disc-valve twin, which by now sported an 8-speed gearbox, but signed the ex-Suzuki teamster Toshio Fujii. He made his Kawasaki debut at the West German GP at Hockenheim, holding an excellent sixth place before retirement.

The next round of the world series was the Isle of Man TT; but after spending much of the practice week awaiting gearbox parts Fujii finally

Kawasaki's A1-R. First seen in 1966, this 'over the counter' racer was powered by a 247 cc disc valve twin-cylinder engine which produced 37 bhp and was good for over 120 mph

Englishman Dave Simmonds with the works 125 twin he rode to eighth place in the 1966 Japanese GP. Simmonds later took the 125 world title in 1969 – his and Kawasaki's first

went out on the Friday evening practice session only to suffer a fatal fall at Crookshanks Corner, Ramsey.

As Fujii was effectively the factory's only GP rider, the 1966 challenge died with him, and it was not until the final classic of the year, in Japan at the Fisco circuit, that Kawasaki reappeared. At Fisco, they had quite a large presence with both 125 twins in the GP, plus 90 and 250 cc models in the Junior clubmans type events. The other news was quite sensational: the existence of a brand new four-cylinder 125 GP mount.

Naomi Tanaguchi, formerly a Honda works rider and development man, had joined the Kawasaki race department. Additionally, Ernst Degner, the former MZ and Suzuki star, was injured when he crashed while trying out a works 125 Kawasaki twin at Fisco prior to the Japanese GP. This was Degner's third serious accident in as many seasons and it was to signal the end of a brilliant, if blighted career.

For the Japanese GP, Kawasaki had also made what was to turn out as an

Above
During the mid-1970s a European team was created with Mick Grant and Barry Ditchburn as riders, and Stan Shenton as team manager. All three are seen in this 1977 photograph, together with their support crew

Above right
Yorkshireman Mick Grant in vivid action on one of the works KR250 in-line twins in April 1978

Right
The KR250 in-line twin. It employed disc valve induction. The 249.1 cc (54 × 54.4 mm) motor produced in excess of 60 bhp at 12,500 rpm. Other notable features included electronic ignition, dry clutch and 6-speed gearbox

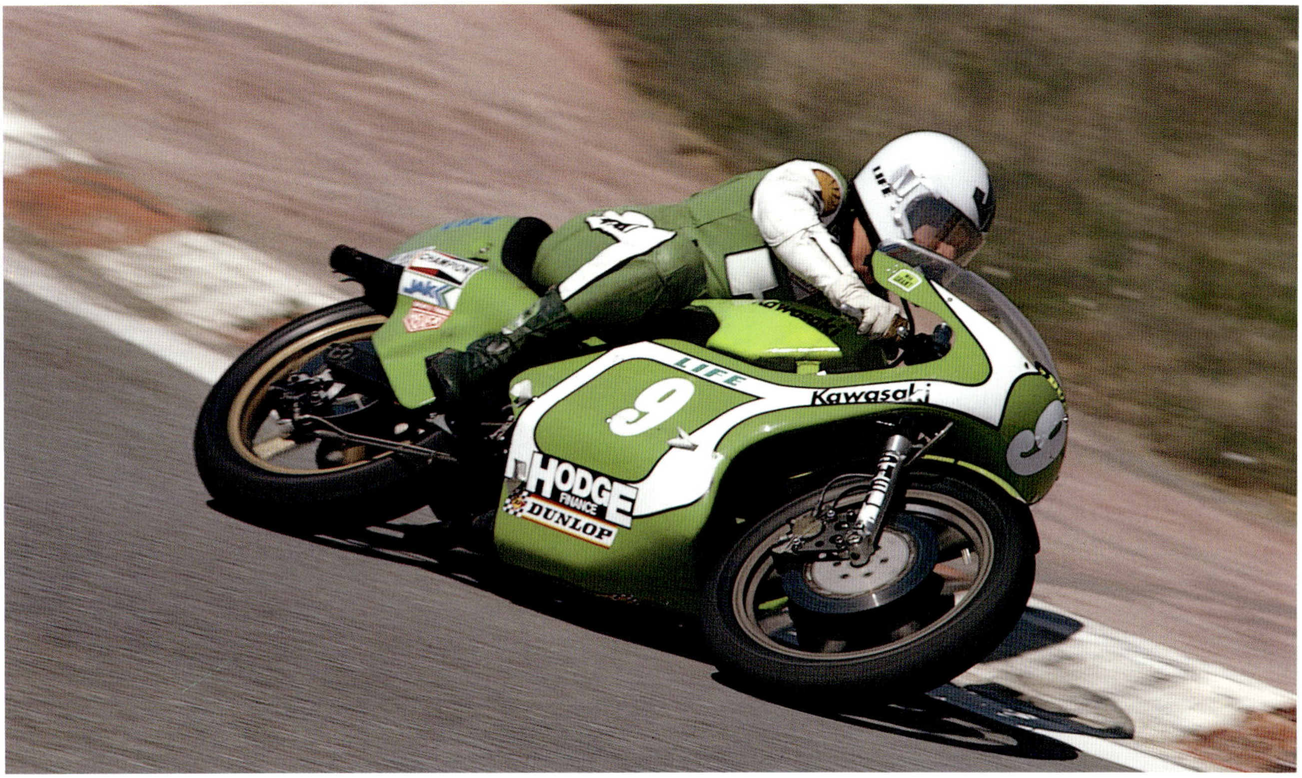
LIFE
9
Kawasaki
HODGE
FINANCE
DUNLOP
CHAMPION

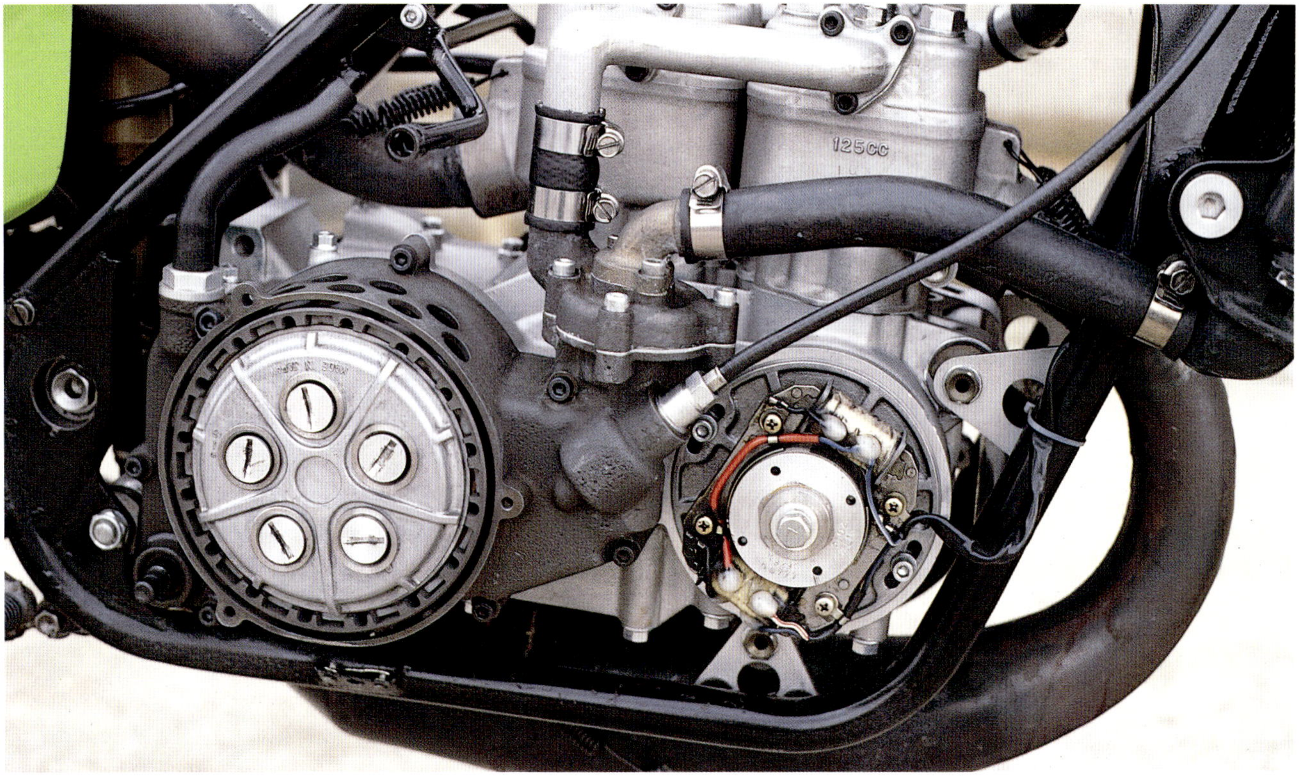
125CC

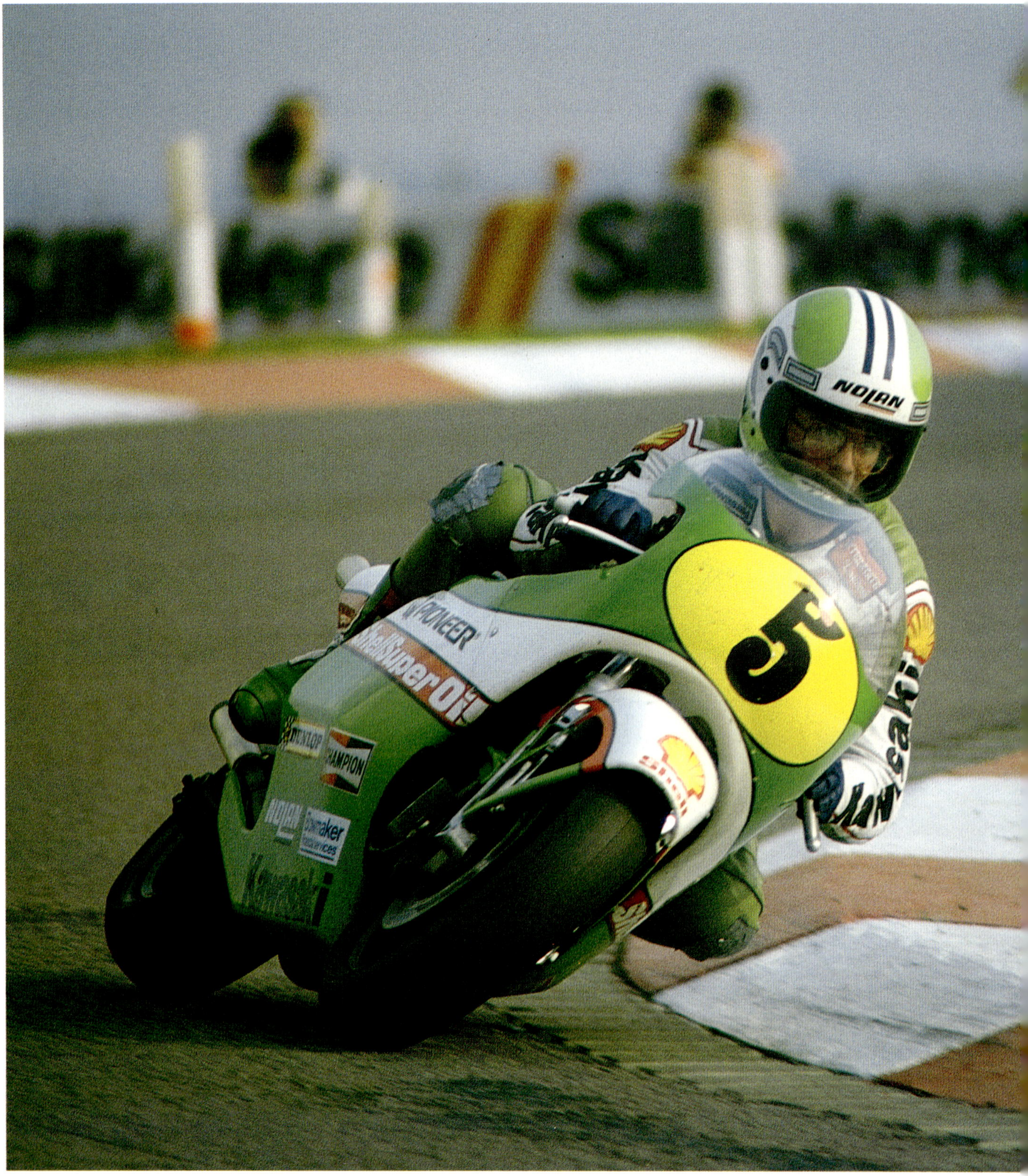
NOLAN
5
PIONEER
CHAMPION
DUNLOP

inspired move by gaining the services of the Tohatsu rider Dave Simmonds. Simmonds finished eighth, the first Kawasaki home, resulting in a full contract for the following season.

The clubmans type racers consisted of a 90 cc racing version of the top-selling Kawasaki roadster, whilst the 250 was a competition version of the shortly to be released Samurai disc-valve, twin cylinder super sports roadster. Coded A1-R this featured a 5-speed gearbox and lubrication by a crankcase-mounted pump supplying oil to the main bearings and big-ends. Performance, if not reliability, was on a par with the latest Yamaha TD1 production racer.

Although the 125 four was raced in the Singapore Grand Prix (by Japanese riders) it was never destined to reach Europe. Instead, Simmonds was provided with the latest version of the twin; but generally 1967 was to prove a poor year on the GP trail for Kawasaki. Simmonds also suffered a serious crash at the Ulster which was to sideline him for almost all of the following season.

Into 1969 and the Englishman still had the loan of his by now ageing works 125 twin. With Suzuki and Yamaha officially quitting the sport, Simmonds secretly thought that he 'just might' at last stand a chance of that elusive grand prix victory. However, in case the term 'works rider' conjures up the image of a group of factory mechanics in beautifully laundered overalls, and a motorhome, just ponder this: Dave Simmonds was very much on his own – one man riding and working on his Kawasaki, his base a caravan-cum-workshop lit by bottled gas.

If the truth be known, before the 1969 season began, the factory had given up all hope of not only winning a world championship, but also of even securing a single GP victory.

In fact, prior to the season getting underway, as Simmonds remarked afterwards: 'They sent me a box containing two spare crankshafts, four pistons, a few sets of piston rings and various other oddments. It was virtually all they had left in stock. That was it, and the best of luck!'

With this in mind, what actually happened is all the more amazing. Over an 11-round championship, Simmonds won eight rounds, and this after missing the first in Spain through lack of start money!

However, if the above gives the impression that his Kawasaki ran faultlessly, nothing could be further from the truth. Almost from the start, Simmonds suffered endless problems during practice, qualified with the barest minimum of laps, and then had a virtually trouble-free race

The South African rider Kork Ballington shot to instant fame by taking the 1978 250 and 350 cc world titles on the latest versions of the in-line twins. 'Korky' is seen here on a KR500 at Donington Park, 1981

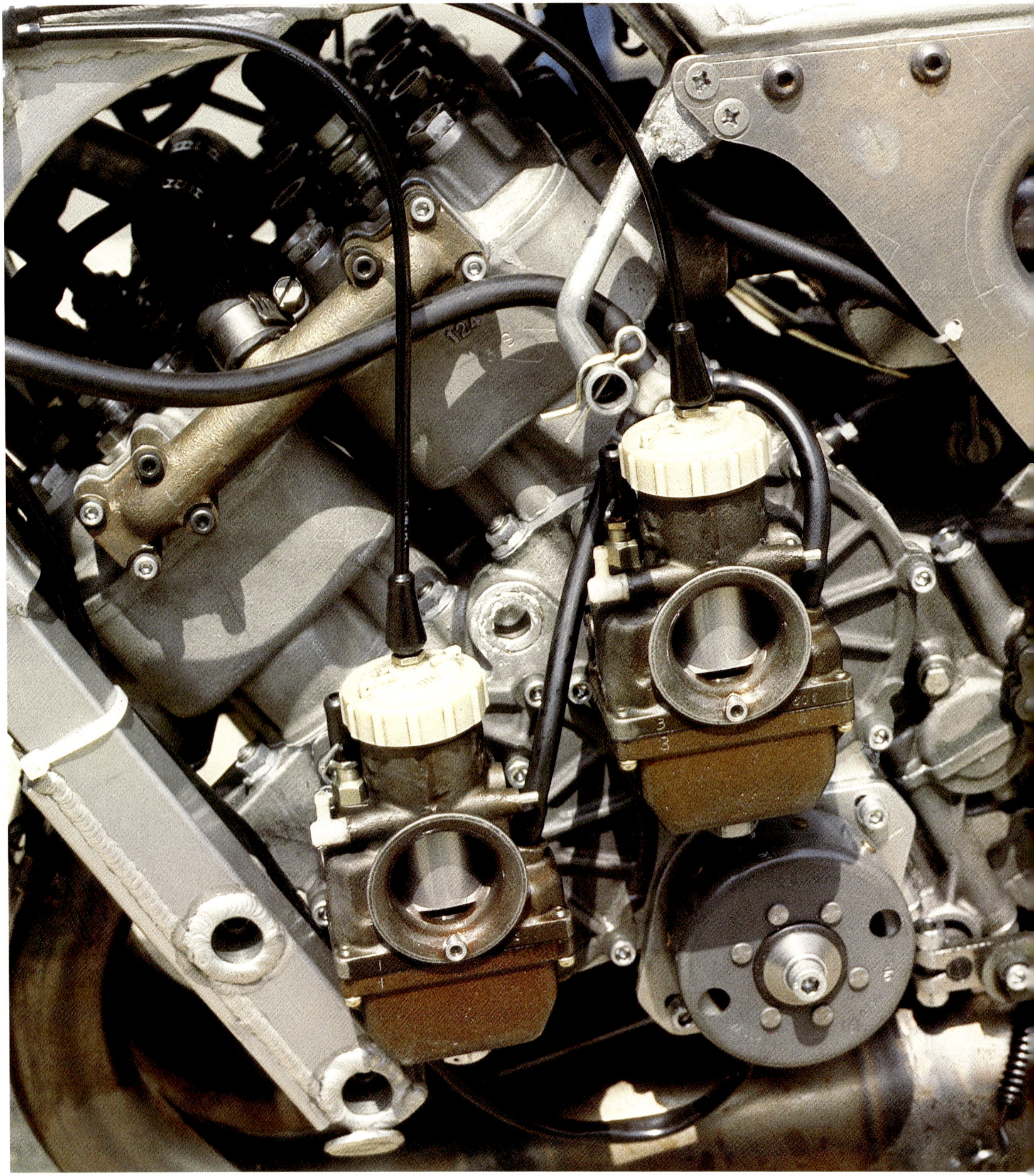

proper after more often than not getting away in mid-field.

After all the glory of winning the title, what followed was very much an anti-climax, for Dave took the bike to the non-championship Austrian GP and well and truly wrote it off by slamming into a wall!

The three-cylinder Mach III 500 cc sports roadster had been launched in 1969 and that year saw a succession of honours in Production Sports machine races. With results such as this, it came as no surprise when Kawasaki entered the 500 cc road-racing class in 1970 with a pukka racing version, the fearsome, firebreathing H1-R.

The H1-R used the piston-port induction and general design of the three-cylinder roadster. Its 498 cc (60 × 58.8 mm) engine also employed the usual Kawasaki lubrication system, which injected oil into the inlet tracts from an independent supply, but, in addition, the fuel was a 20:1 petrol-oil mixture.

Each of the three barrels and heads were separate castings, with two-ring pistons. A cross-over layout of exhaust pipes was used to reduce the frontal area and to tuck the expansion chambers neatly out of the way. Carburation was taken care of by a trio of 34 mm Mikuni instruments. Instead of the electronic ignition of the roadster, the racer had an alternator, rectifier, battery and three ignition coils. The rectifier and coils were mounted on rubber from a bracket projecting forward from the steering head. Maximum rpm was 9000, with over 70 bhp on tap. Like the roadster, the gearbox was a 5-speeder, but the clutch, which ran on the offside, was dry and exposed.

The potential of the new 500 triple was immediately evident, and at Daytona in March, Rusty Bradley won the 100-mile 'amateur' race in convincing fashion. His average speed of 100.72 mph beat the record set the previous year by a full 4 mph.

Daytona also showed the H1-R's major fault for longer races – an almost unbelievable fuel consumption of 11 mpg!

The most successful H1-R in GPs that year was the New Zealander Ginger Molloy, who finished the season second only to Agostini and MV Agusta in the points table. Molloy's achievements included runner-up spots in France, Finland, Ulster and Spain.

Simmonds rode an H1-R that year, but with little success, while his rebuilt 125 came home third in the title hunt. His last major victory on the machine came in the West German GP in May 1971.

First seen in 1980, the monocoque-framed KR500 was not the success for which the factory hoped; its highest ever positions in grand prix were a couple of third places in Holland and Finland during 1981

Kawasaki
GOODYEAR
21

PIRELLI
RIZLA+
TEAM GREEN
D.I.D Racing Chain
NGK SPARK PLUGS
Kawasaki
Arai

Left
A previously unpublished shot of Eddie Lawson with the KR500 he rode at Daytona in 1982; this was one of the very last appearances for Kawasaki's two-stroke four before the company switched its racing effort to four-strokes

Below left
After several seasons of reliable, rather than outstanding performances, John Reynolds emerged as the man to beat in British Superbike racing during 1992

Although Dave Simmonds continued racing Kawasakis, he was to lose his life tragically during the caravan fire of his friend and fellow racer Jack Findlay on the eve of the 1972 French GP.

Meanwhile, a new 750 had appeared which was to have a significant bearing on Kawasaki's future direction in racing circles. Its first year saw a mass of reliability problems, notably at Daytona, but once these had been overcome it started to show real potential.

In 1973 the 750 H2-R triples won just about everything in America — except Daytona, where once again luck deserted them.

Like the Mach III before it, the introduction of the 900 Z1 saw a whole clutch of successes in Sports Production class racing. Fired up by this, Kawasaki went endurance racing with a vengeance in 1974, using specially tuned Z1 engines in an Elgi chassis'. Their top pairing of Frenchmen Georges Godier and Alain Genoud won at Barcelona, Mettet and the Bol d'Or at Le Mans.

If 1974 was good, 1975 was even better. Godier and Genoud became World Endurance Champions, while the new water-cooled version of the 750 two-stroke triples gained considerable success, ridden by the likes of Yvon Du Hammel, Barry Ditchburn and Mick Grant. The latter pair having been recruited into a new European team run by Stan Shenton. The highlight of the year was Grant's Senior TT victory. The Yorkshireman also broke Mike Hailwood's long standing 1967 absolute lap record for the 37-mile Mountain circuit at 109.82 mph.

By 1976 a specially built Daytona version of the 750 triple, using a 6-speed box and giving 115 bhp was claimed to be capable of 185 mph. Unfortunately for Kawasaki even this mighty machine couldn't beat the likes of Kenny Roberts and Johnny Cecotto that year.

Back in 1975 Grant, together with other riders, had race tested a brand new 249.1 cc (54 × 54.4 mm) tandem twin. This disc-valve engined device didn't show its true potential until some two years later, when Mick Grant scored the company's first ever GP victory in the 250 cc category at the Dutch TT. Also after a poor TT in 1976 the same rider stormed back to win the 1977 Classic TT at a record average speed of 110.76 mph and also set a new absolute lap record of 112.76 mph.

The year 1978 was, without doubt, Kawasaki's greatest ever season on the tarmac. American Don Vesco raised the motorcycle world speed record to 318.598 mph using a fully streamlined projectile powered by a pair of turbocharged Z1 engines. On the race circuit new signing Kork Ballington shot to instant fame by taking both the 250 and 350 cc world titles on the latest versions of the in-line twins, while Mick Grant continued his run of success in the Isle of Man by not only winning, but again pushing up the lap record, this time to a staggering 114.3 mph.

Although Grant subsequently quit to join Honda for 1979, Kork

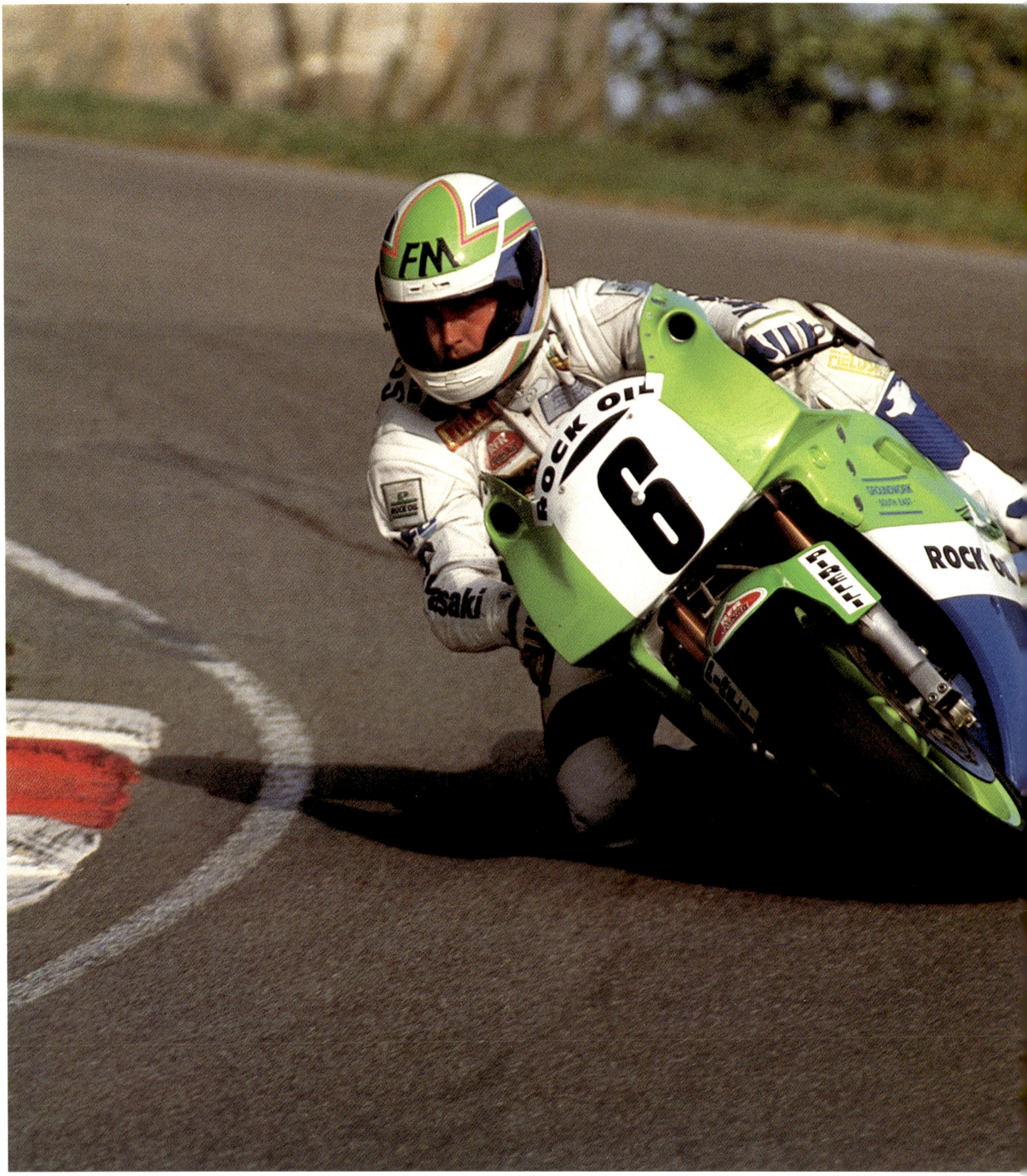
FM
ROCK OIL
6

Ballington continued, gaining another 250/350 cc world championship double.

Into the 1980s and the Kawasaki effort became 'Team Green'. There was more world championship success: West German Anton Mang taking the 1980 250 cc world title, 1981 250/350 cc championships and 1982 350 cc. As for Ballington, he had been drafted onto the new KR500 four-cylinder two-stroke; unfortunately this machine was never to reap the success garnered by its smaller brothers.

At the end of 1982 Kawasaki announced its retirement from grand prix racing. This was for several reasons, most notably the downturn in sales worldwide and the fact that the company was selling mainly four-strokes, while its GP racers were two-strokes. During the eight years of grand prix glory (1975-1982), Kawasaki had developed several innovations, including Uni-Trak rear suspension, air-adjustable front suspension, and Electro Fusion cylinder bores.

From 1983 onwards the Team Green racing effort was placed very firmly on what was to emerge as Superbike racing. Leading contenders have included Eddie Lawson, Wayne Rainey, Rob Phillis and John Reynolds. As results have proved, when it comes to large capacity four-strokes, Kawasaki can produce machines to match the best in the world, with their combination of lightning speed and awesome reliability.

An ex-schoolboy motocrosser, John Reynolds is shown heading for yet another victory with his ZXR750R in the summer of 1992

Four-stroke twins

Although Kawasaki had offered their own version of the BSA conceived/Meguro copied 650 ohv twin in the late-1960s, the first 'all-Kawasaki' four-stroke twin was the Z400 of 1974.

At its heart was a single overhead cam parallel twin of 398 cc (64 × 62 mm); other features of its technical specification included a compression ratio of 9.4:1, twin 36 mm Keihan carbs, Hy-Vo chain primary drive, 5-speed gearbox, 12-volt electrics, single disc front brake, lockable fuel cap, audible turn signals and two helmet locks. Maximum speed was a shade under 100 mph, while fuel consumption averaged around 50 mpg; even ridden against the stop it would achieve 45 mpg.

The Z400 was no sportster. Instead it was designed as an all-function tourer. Compared to other 400s of its day – CB400/4, KH400, GT380 and RD400, it lost out on both top speed and acceleration, yet it still won friends and customers as a maid of all work. Perhaps *Bike* magazine summed it up best in a May 1978 test – 'It accomplished everything with efficiency, but without inspiring the rider to try harder'.

Except for styling updates the Z400 remained virtually unaltered throughout its long life. The exhaust was one of the few changes and then

Left
Although Kawasaki offered its own version of the BSA-conceived, Meguro-copied 650 ohv twin in the late 1960s, the first 'all-Kawasaki' four-stroke twin was the Z400 of 1974

Right
The year 1977 saw the launch of the Z750 – the only mass production, large capacity twin at the time to sport double overhead camshafts. Sales never matched expectations

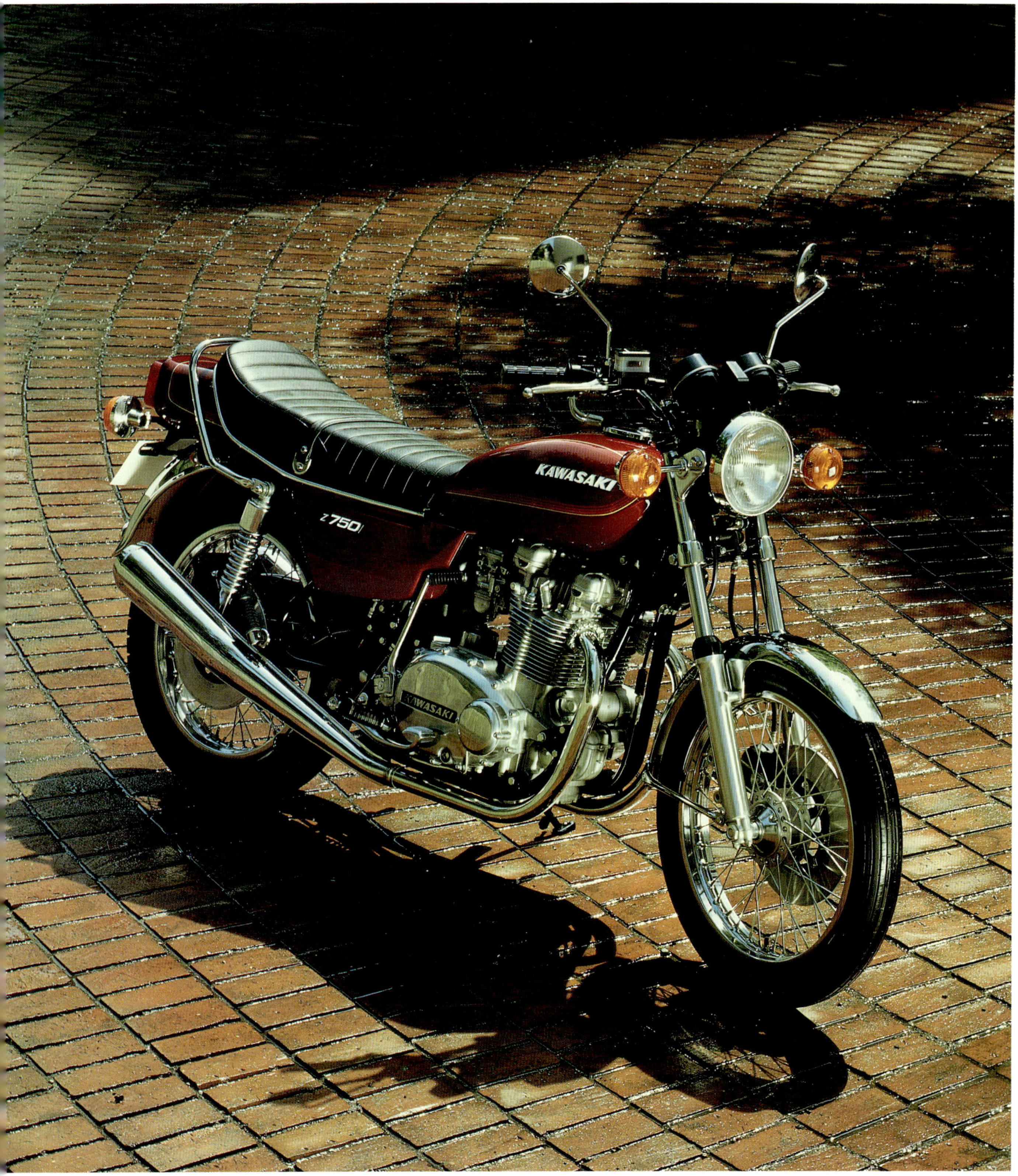
KAWASAKI
Z750
KAWASAKI

Left

The Z250 made its British debut in the 1980 model year. It was built to compete with similar bikes from the other 'big three' Japanese manufacturers: Honda (Super Dream), Suzuki (GSX250) and Yamaha (XS250)

Right

During the late 1970s the original Z400 was enlarged to 440 cc; this is the first of the larger engined models, the 1980 C1. Although, like its predecessor, a reliable workhorse it was still largely a bland and boring bike to ride

it was only made to conform with projected US emission controls.

The Z400 was originally introduced as the company's alternative to the exciting but fuel-gobbling three-cylinder 400S3/KH400 two-stroke. The contrast couldn't have been more noticeable! Instead of the ultra-sporting temperamental two-stroke, the Z400 was bland, but ultra-reliable. Perhaps its only lasting claim-to-fame is that it was one of the earliest Japanese bikes to be fitted with rotating balancer shafts as a means of dampening vibration. In 1980 the Z400 became the Z440 by the simple expedient of boring the cylinders out to 67.5 mm; this increased the capacity to 443 cc.

If the Z400/440 series could be described as bland but good sellers, the next of the Kawasaki four-stroke twins, the Z750, was not only bland but a real sales flop.

On paper it looked good, but in practice potential customers just were not interested, plumping instead for one of the many four-cylinder 750s on offer elsewhere.

Kawasaki had seen their big twin as a modern update of a design which still had a warm place in the hearts of motorcyclists who longed for the simplicity and character of the old bikes, but wanted one in a modern engineering package.

At the time of its launch (1977) the Z750 was the only large capacity twin with double camshafts. In theory this meant better performance; Kawasaki claimed the 745 cc (78 × 78 mm) engine to be 'as smooth as a big four'. They also claimed that there was 'bags of torque' and 'surprising economy of fuel'. Maybe there was, but ultimately customers, or the lack

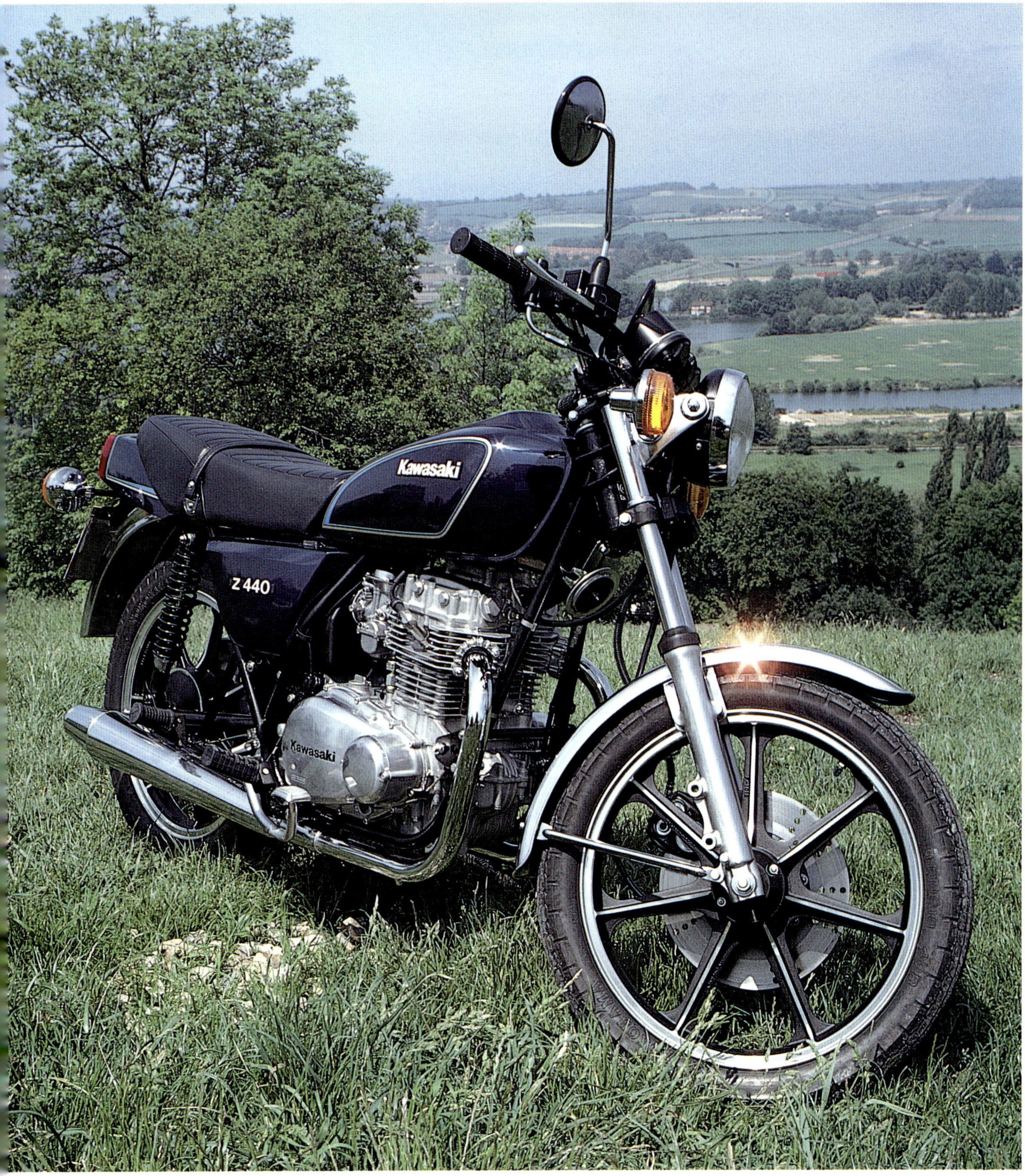
Kawasaki
Z 440
Kawasaki

of them, determine the success or failure of any motorcycle. No one seemed to want the Z750 and so it quickly died (a custom bike using basically the same motor appeared later, or was this maybe a case of using the existing stocks meant for the Z750 roadster?).

Next came a completely new venture: a high performance, lightweight sports 250, capable of keeping pace on the road with bigger machinery, but with economical fuel consumption and precise, crisp handling. The result of this design exercise was to be the Z250 (Scorpion) which made its British debut in the 1980 model year. At that time each of the other three Japanese manufacturers had a similar machine on offer – Honda CB250 Super Dream, Suzuki GSX250 and Yamaha XS250.

Kawasaki's attempt was perhaps the best of all these bikes, its 248 cc (55 × 52.4 mm) producing 27 bhp (1 bhp more than the market leaders Honda) with good torque and a maximum speed of 97 mph. There weren't four valves per cylinder, as on the Super Dream, but the motor could still spin over at 10,000 rpm, enabling the bike to cover 400 metres (a quarter mile) in 15.5 seconds from a standing start. Other features included electric start, front and rear disc brakes, a 6-speed gearbox and a dry weight of only 153 kg (337 lb) – 10 kg less than the Honda design. Subsequent versions saw the introduction of Uni-Trak rear suspension, tubeless tyres and even belt final drive.

From the Z250 came the Z305 (and later GPZ305). The larger capacity was achieved by boring the cylinders out to 61 mm. Maximum power increased to 36 bhp at 10,000 rpm and maximum speed to 105 mph.

The 305 has remained virtually unchanged (except for the addition of belt drive for the 1983 season) and was still listed (in B8 guise) when this book was being written in the autumn of 1992. British price in 1992 was £2410 against £1249 (including taxes) when launched in 1982 – this provides some idea of inflation in bike prices over a ten year period – virtually a 100 per cent increase for what amounts to the same machine.

Until now all the four-stroke twins described have been somewhat unexciting creations, at least for sports bike enthusiasts. This all changed in 1987 with the arrival of the GPZ500S (EX500 in certain markets) with its liquid-cooled dohc 8-valve motor.

At the time of its launch the GPZ500S introduced a brand new concept to the motorcycle industry – a budget priced sports bike with slim lines, hi-tech engine and low weight, 169 kg (372 lb) dry.

Developed from the Z250, the 1983 EX305 (GPZ in UK) offered riders the chance to own a budget-priced sportster; it has sold well ever since. Belt final drive comes as standard equipment

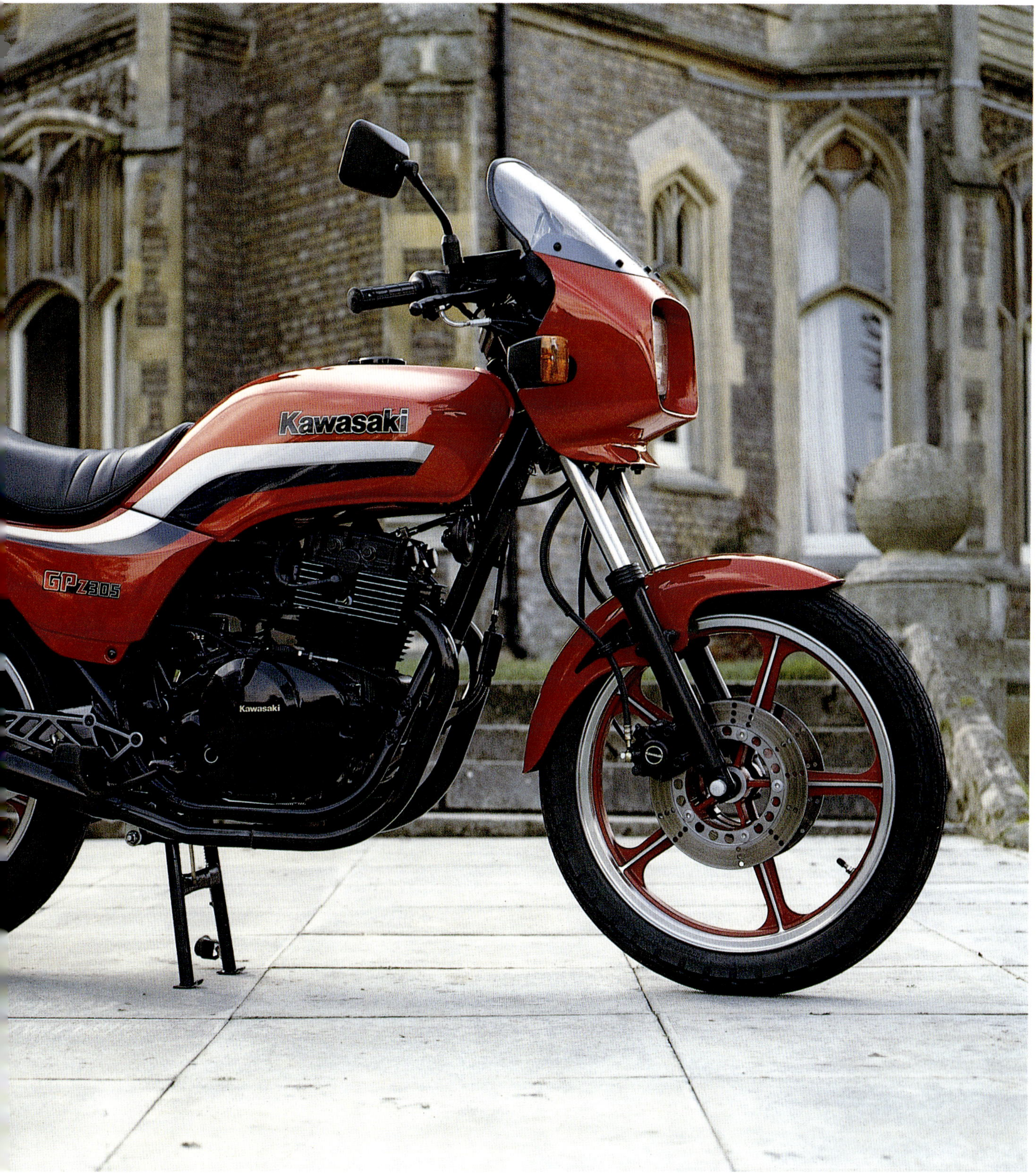
Kawasaki
GPz305
Kawasaki

The 498 cc (74 × 58 mm) dohc 8-valve with liquid-cooling was equipped with a gear-driven counterbalance shaft plus rubber engine mounts at the front to dampen out vibration. A similar engine (based on half a GPZ900R) already powered the EN450 custom bike, but the heart of the GPZ500S was based on the GPZ1000 RX engine, giving 60 bhp at 9800 rpm.

The frame was equally up to date, being a perimeter type made from box section high tensile steel, and utilising the bottom-link Uni-Trak rising rate rear suspension. The offside down tube unbolted for easy engine access, otherwise the frame was of all-welded construction.

The 280 mm single front disc utilised the same caliper that was already fitted to the GPX750R, employing BAC (Balanced Actuation Caliper). In this there were two different sized pistons used to grip the disc, balancing, Kawasaki claimed, braking force and spreading it evenly across the pad for less grab, better feel and longer life.

As standard the GPZ500S was fitted with a half fairing which mounted flush to the 18-litre fuel tank. A belly pan was an optional extra; 16-inch wheels completed the specification.

Maximum power was 60 bhp at 9800 rpm giving a top speed of 128 mph, thanks in no small part to its excellent power-to-weight ratio.

Both the 305 and 500 remain in production, and both have seen virtually no changes throughout their life; a fitting tribute to an important sector of Kawasaki's model range, the budget priced four-stroke twins.

Since its introduction the GPZ500S has remained largely unchanged except for colours and graphics. This is the 1991 A5 model. Although the engine tends to be a revvy unit without much mid-range torque, performance and handling are first class

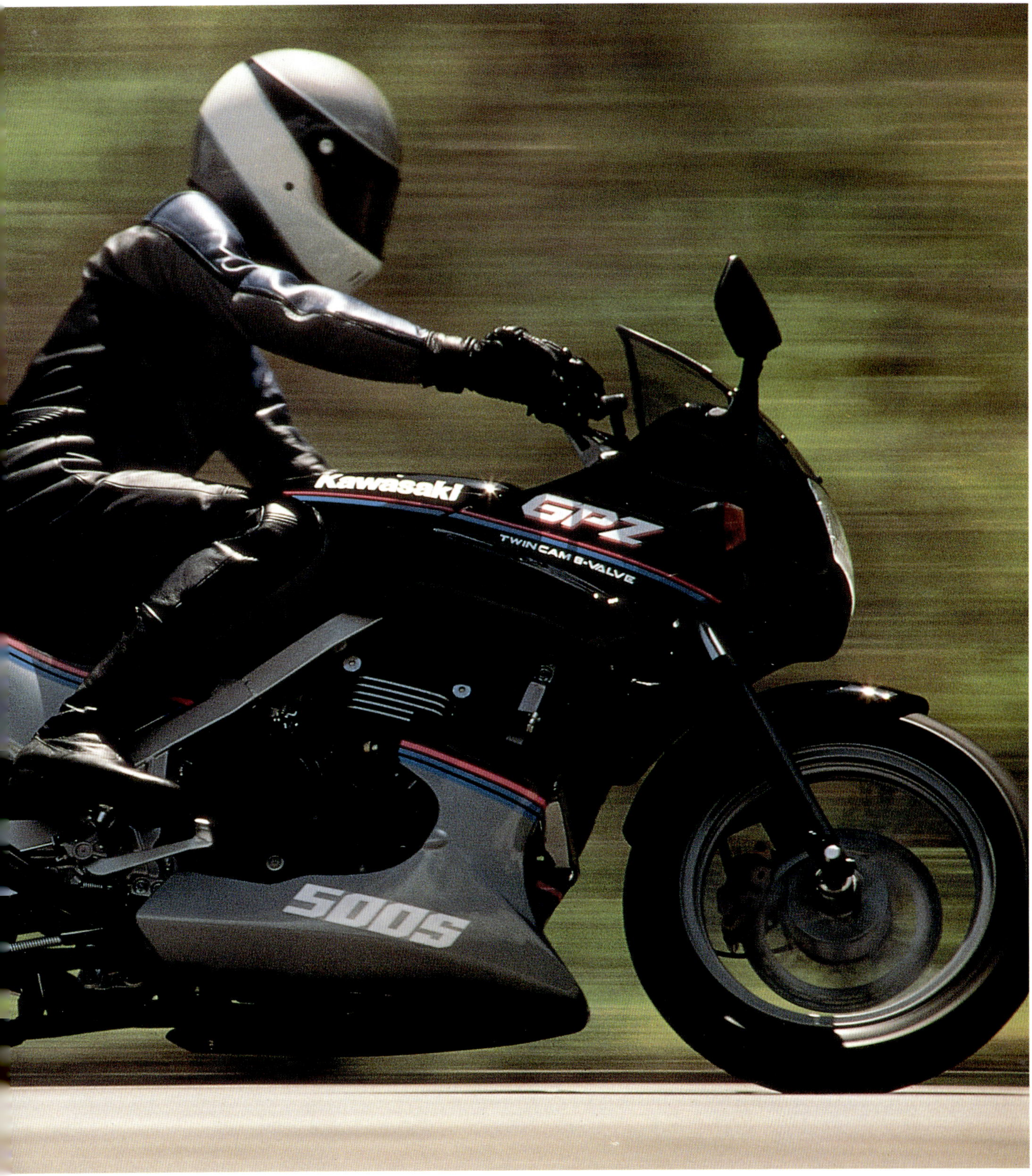
Kawasaki
GPZ
TWIN CAM 8-VALVE
500S

Z1300

The mighty six-cylinder Z1300 ran for almost a decade. In no way could it be called pretty, but none the less its almost brutal appearance tended to mask what was in fact a notably sophisticated piece of engineering. Even the engine layout, behind an enormous radiator, did nothing to enhance it aesthetic qualities. The sheer size of the Z1300 is emphasised by its 294 kg (653 lb) dry weight, which is only beaten by Honda's Gold Wing and Harley's Electra Glide; however both these bikes included comprehensive fairings and luggage systems; the Z1300 was all naked muscle. The dimensions 1580 mm (62.2 in) wheelbase and seat height of 835 mm (32.3 in), provide the evidence, if any were required, that this was a big motorcycle.

One person who gives the reader a good idea of just what you could or could not get away with on a Z1300 is Arto Nyqvist, a stunt rider from Helsinki who proceeded to amaze crowds throughout Europe with his antics, both on and off the bike. He pulled everlasting wheelies as if he were on a dirt bike and his speciality was to get his Z1300 up to 100 mph, then step off and be pulled along behind. This provides proof of the inherent stability of the Z1300, if only in a straight line, although it must be said that Arto handled his machine with great skill even on the turns.

Road testers' opinions of the Kawasaki giant seemed to differ very little, which is quite unusual. On testing the Z1300, most appeared surprised at how easy it was to control and found that it was not the bucket of sludge which they had expected.

Although the Z1300 was very much a sports tourer, it would quite happily pull from as low as 800 rpm in top to its 8000 rpm red line with virtually no hesitation or strain at all. The combination of six pots and shaft drive provided not only a unique, but very smooth and rapid rate of travel. Add to this a reasonable level of handling and braking allied to supreme comfort, and one had an excellent motorcycle for travelling long distances – the only real failing being the amount of wind buffeting due to the lack of a fairing.

The Z1300 was launched in December 1978, going on sale in January 1979. Production lasted some ten years and ended in 1988. There were

The truly awe-inspiring six-cylinder Z1300 was launched in December 1978 and remained in production for almost a decade before being finally axed in mid-1988; this is the original 1979 A1 model

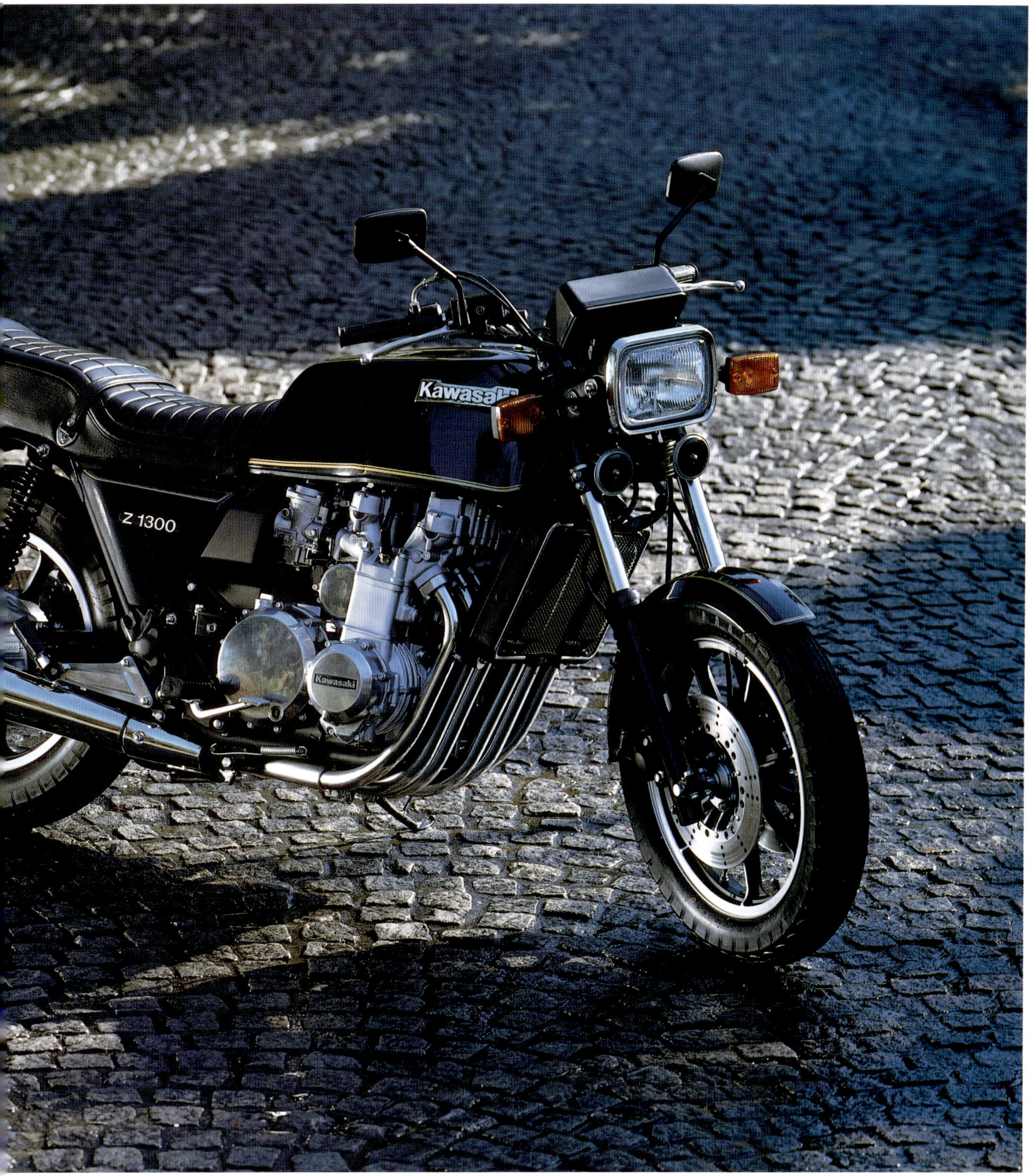
Kawasaki
Z 1300
Kawasaki

Above
Z1300 crankshaft and transmission exposed for all the world to see

Right
At its heart the Z1300 has a liquid-cooled, six-cylinder, dohc transverse engine. Its long-stroke 62 × 71 mm dimensions gave it a capacity of 1286 cc

surprisingly few major changes along the way from the original 1979 Z1300 A1 through to the final ZG 1300 G3 of 1988.

At its heart the Z1300 had a liquid-cooled, six-cylinder, double overhead cam transverse engine. Its bore and stroke dimensions gave it a capacity of 1286 cc. Maximum power output of the original 1979 A1 model was 130 bhp at 8000 rpm, enough to provide a maximum speed of over 130 mph – provided you could hold on! At sub-80mph speeds the combination of tubeless tyres, triple disc brakes, a powerful Halogen headlamp, supple suspension, a vibration-free motor and high-density pre-shaped dual seat for extra comfort added up to make a motorcycle which was eminently suitable for two-up touring.

For the 1980 model year the model prefix was changed to A2, but the only difference was in colour choice. For 1981 the A3 model gained adjustable air rear shocks, while the 1982 A4 featured transistorised ignition with electronic advance. Little change was made for the A5 of 1983, but 1984 saw the Z1300 G1; this had digital fuel ignition (DFI),

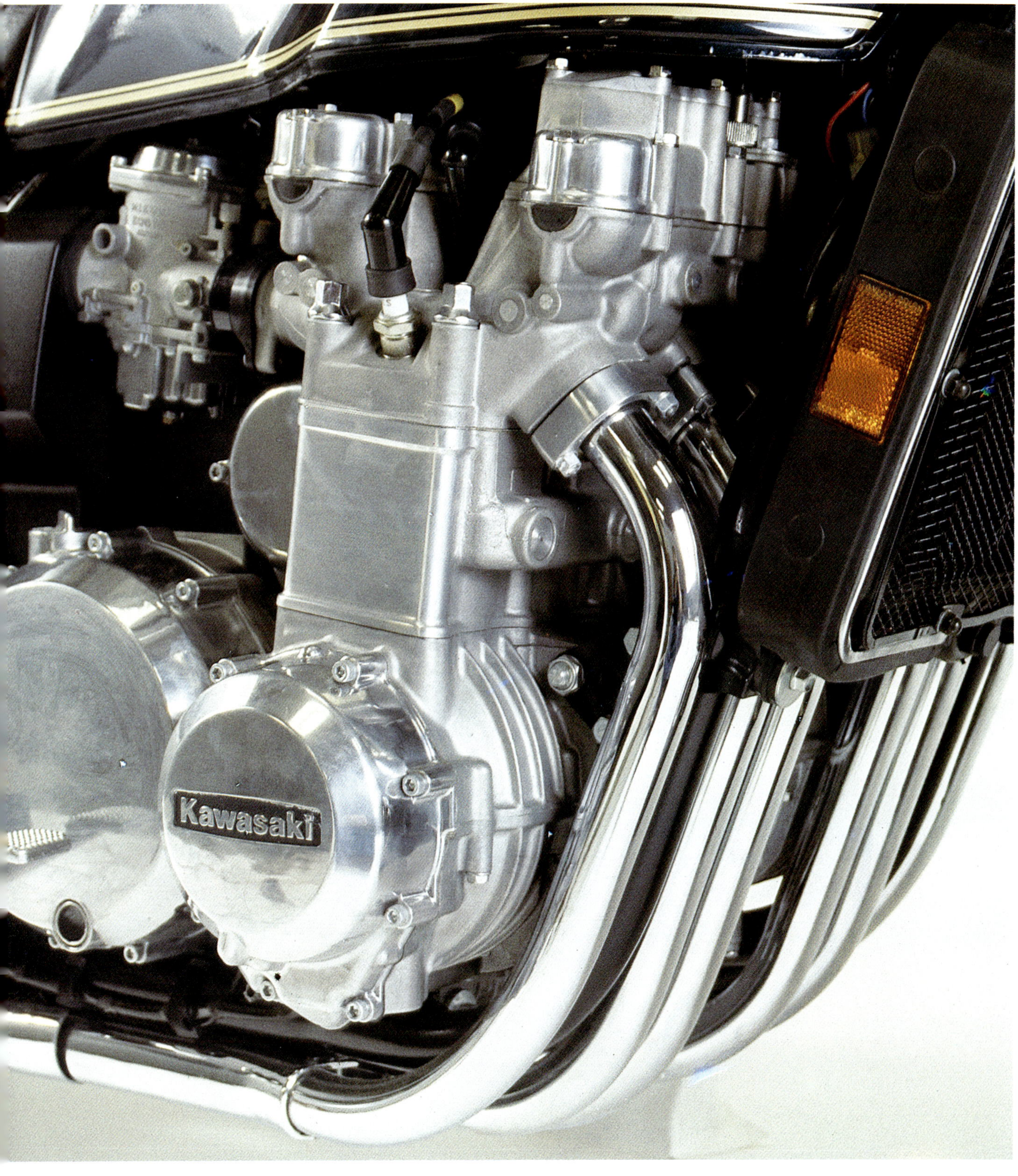
Kawasaki

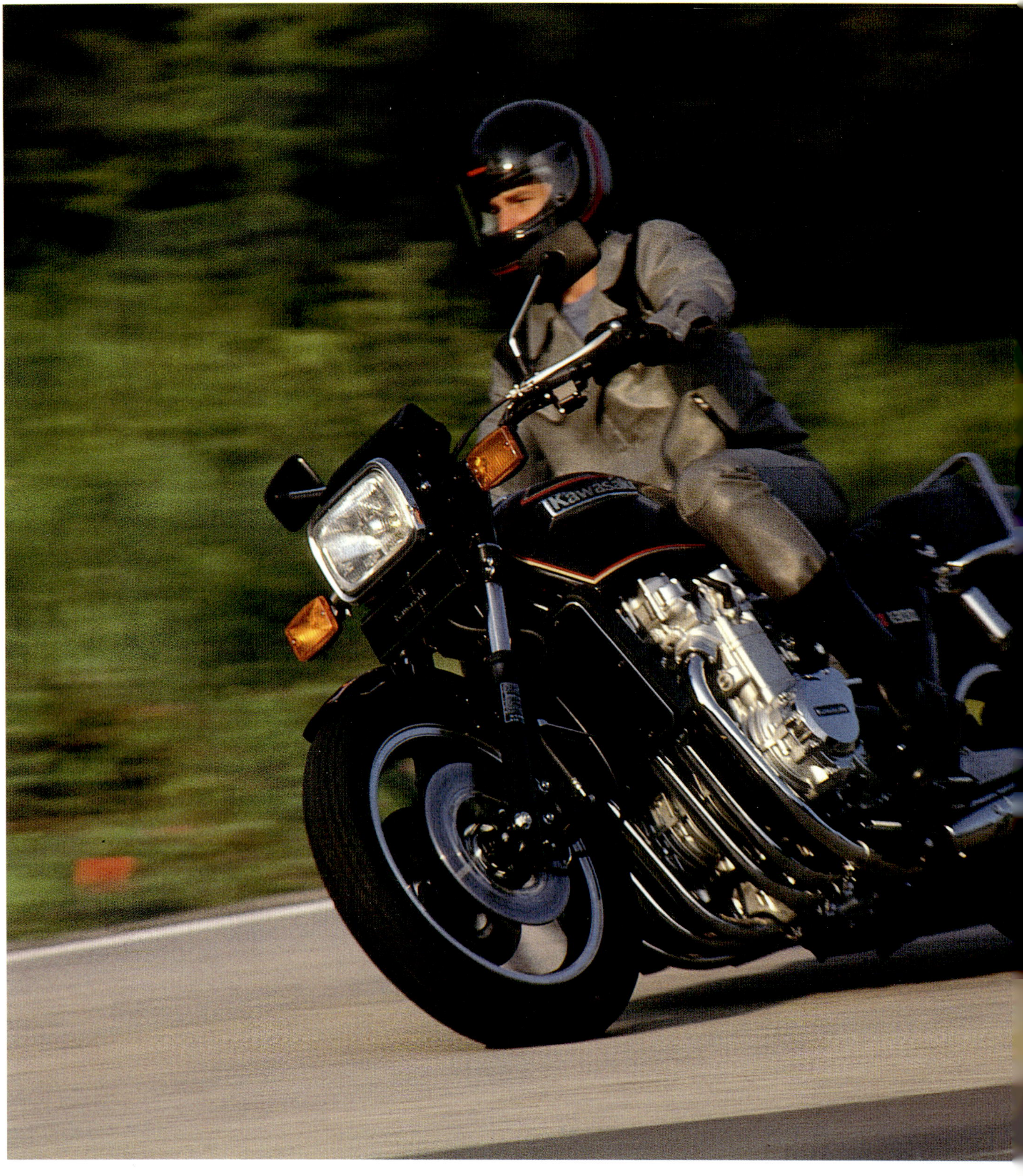

which had been pioneered on the GPZ1100 the previous year. The DFI system replaced the three double choke constant velocity Mikuni carburettors. Compared with the more conventional electronic fuel injection (EFI), the DFI has no flaps or gate to measure the air flow, which leaves the intake passage completely unobstructed and thus free from the turbulence created by such sensors. Even more importantly, it eliminates the slight delay which occurs between the opening of the throttle and the response of the air flap to the increased air flow.

The system eliminates the tendency for the air flap to bounce when the throttle is closed and then immediately snap open again – this can cause a sudden hesitation just when immediate acceleration is most needed. In place of the EFI system's restrictive air sensor, the DFI reads the throttle opening, engine revs, air and engine temperature plus the atmosphere pressure. It then instantaneously computes the appropriate fuel injection rate for optimum performance.

The digital fuel injection system also has a 'fail-safe' circuit which allows the motorcycle to be ridden even if the electrics pack up. Other advantages of the DFI are smoother and more immediate throttle response, easier starting (regardless of engine temperature) and significantly improved high altitude performance (the altitude sensor can even register changes in the barometric pressure due to the weather). Fuel consumption and exhaust emissions are improved.

In 1986 the G1 became the G2, but with no actual changes. The final year of the Z1300 was 1988, by which time the model had changed yet again, to the G3. By this time the Z1300 had been in production for almost a decade, and while motorcycle design and production techniques had gone forward at a truly startling pace, the basic design and styling of the Kawasaki six had remained virtually unchanged. The result was extinction for a bike which, although never produced in large numbers, had none the less created its own niche in the market place as the ultimate heavyweight sports tourer. Only its lack of weather protection as standard equipment spoiled an otherwise clean sheet. Its best description is that of Gentle Giant.

The sheer size of the Z1300 is emphasised by its 294 kg (653 lb) dry weight, which at the time of its launch was only bested by Honda's Gold Wing and H-D's Electra Glide; however both these bikes included comprehensive fairings and luggage systems, whereas the Z1300 (1983 A5 model shown) was all naked muscle

Custom cruisers

Of the Japanese bike builders, Kawasaki was the first to appreciate – and meet – the demand for an off-the-peg custom motorcycle; a machine with all the looks, style and appeal of a chopper, but with the reliability, practicality and competitive price tag that only a major manufacturer can provide. It was also a route which Harley-Davidson had been following, and still follows to the present day, with considerable success.

That was back in 1975, when Kawasaki's American subsidiary took stock Z900s, fitted a number of customising parts, and produced a limited number of such models for sale.

The 'limited' tag stuck, and soon Kawasaki's factory in Japan was producing their own 'LTD' ranges, featuring a family likeness in style.

The first such machines reached Britain in 1979 and by 1981 the range included the Z250LTD single and Z1000LTD four. There were also custom versions of the Z750 four and Z750 twin.

Another popular model was the LTD440 which received the additional benefit of a toothed belt final drive in 1982.

In this machine, Kawasaki's original 'LTD' styling was personified: pull-back bars, stepped saddle, megaphone mufflers, small diameter/wide section rear tyre, small gas tank and an abundance of shiny paintwork and bright chrome-plate.

The LTD440 Belt-Drive produced 40 bhp; there was relatively gentle vibration for a parallel twin, thanks to a balancer shaft. From testing an example for *Motorcycle Enthusiast* the author can say how impressive the machine was as a conventional roadster, with excellent torque and acceptable handling and roadholding abilities.

Kawasaki grabbed the headlines at the 1984 West German Cologne Motorcycle Show when they unveiled a brand new middleweight custom bike, the EN450. This was really a new machine, owing literally nothing to the LTD440. And compared to the earlier bike, it was radical: radical that is in terms of styling, engineering and performance.

Customers now had a choice between the traditional custom approach of the 440 and the bold (read brash!) approach of the 450.

The first Kawasaki custom bikes were stock Z900s modified by the company's American subsidiary; these were soon followed by a whole host of models including the Z250LTD of 1980

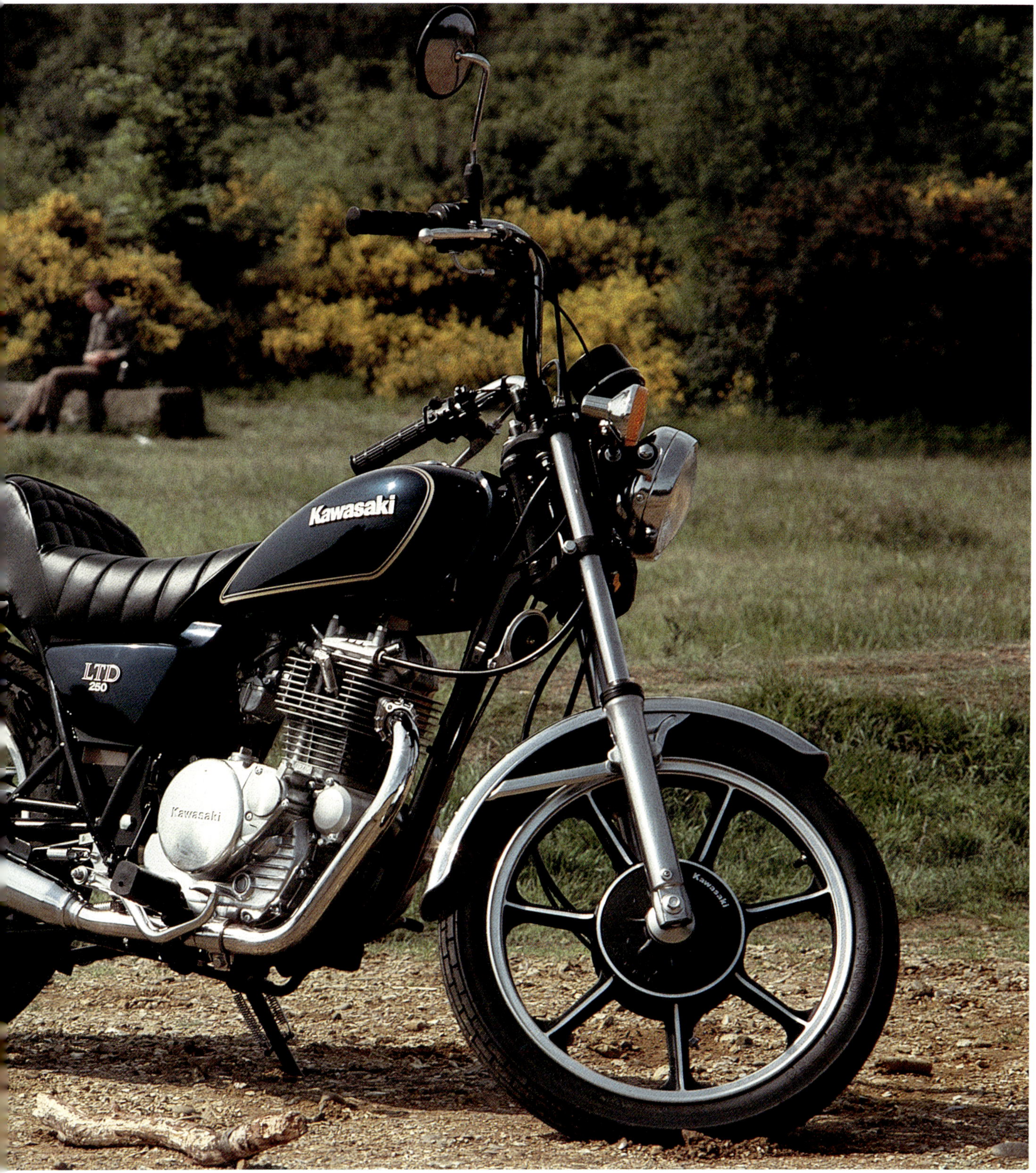
Kawasaki
LTD
250
Kawasaki
Kawasaki

Above
By 1985 when this EN450A1 appeared, Kawasaki were at last building pukka custom bikes rather than lightly converted roadsters

Right
ZL900 Eliminator of 1985: this beefy custom cruiser never made it to Britain, but sold well in the States

Not only had the latter's front forks been given greatly increased rake, but the tank now sported the classic 'teardrop' style to give the newcomer the required long 'n lean look.

Blending in with the tank was the deeply stepped 'King and Queen' seat, which helped provide a really laid back riding pose with the forward mounted footpegs. The rear shocks were laid down, and a distinctive, polished, cast alloy grab rail followed the curve of the rear (shorty) mudguard. The front mudguard was equally abbreviated, the exhausts ended with a stylish slash, and there was extra chrome to be found around the headlamp and instrument pods.

Another feature not found on the LTD440 was the radiator nestling on the frame's twin down tubes. Not a styling exercise, but instead a clue to the parentage of the new powerplant.

While retaining a vertical twin four-stroke layout, Kawasaki had produced an all-new unit which featured liquid cooling, 8-valves, double

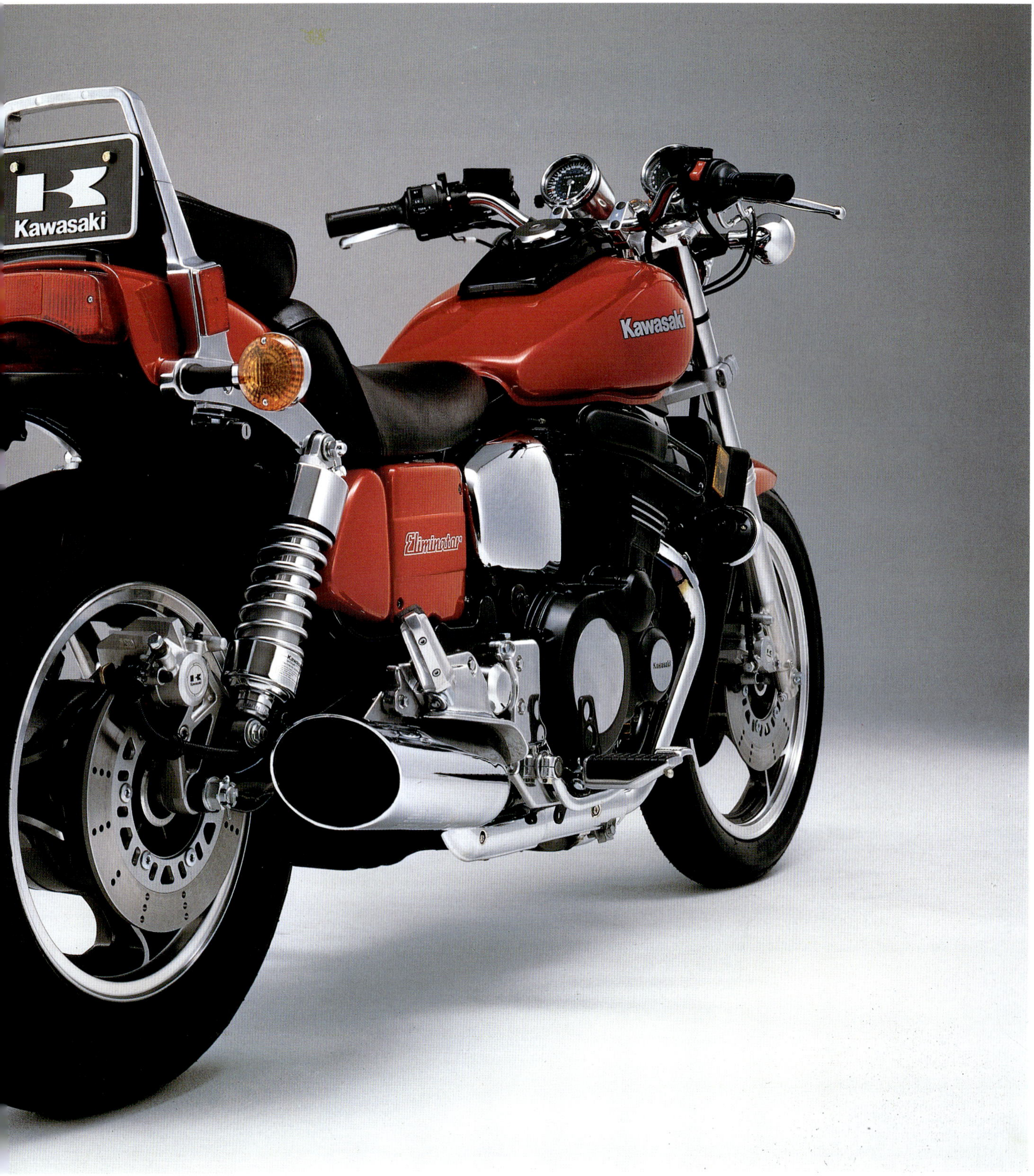
Kawasaki
Kawasaki
Eliminator

Above
Belt final drive of the EN450 was borrowed from the earlier 305 and 440 models; together with the excellent engine it was the best feature of the entire bike

Right
The mighty VN1500 bested even Harley-Davidson in the V-twin capacity stakes, even if it didn't win the custom sales war

overhead camshafts and semi-flat slide carbs. Sounds familiar? It should, because this motor was essentially half a GPZ900R.

The mid-1980s saw yet more Kawasaki customs arrive on the scene, including the VN750 and VN1500 (the latter named Sumo in some markets), V-twins and the ZL900 Eliminator four.

If all this wasn't enough 1987 saw the ZL1000, of which *Cycle* magazine enthused: 'The new King Grunt of roll-on power, the ZL1000 delivers a mid-range punch that crushes the V-Max and snuffs the FJ1200 dead in its tracks'.

This performance was achieved by using a 1000GTR top end crankshaft with a modified GPZ900R crankcase and gearbox. Transmission was via shaft drive whilst the overall gearing was lowered from the other Kawasaki 1000 models so that the 110 bhp output could provide exhilarating straight-line performance.

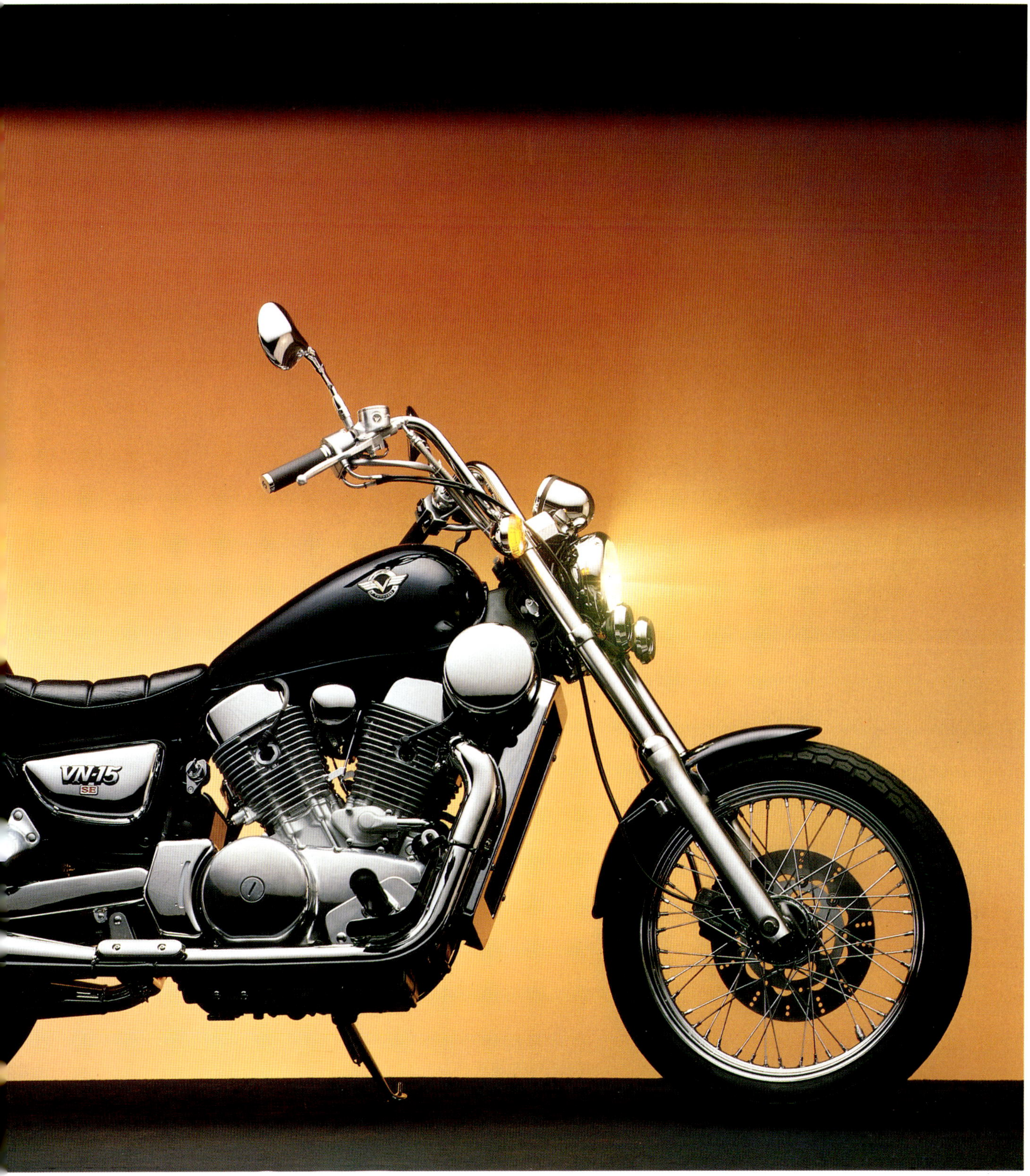
VN-15
SB

The same year the twin-cylinder GPZ500S (using half a 1000RX engine) was introduced. The same powerplant eventually (1990) found its way into the middleweight custom model and the EN500 was born. Since then the bike Kawasaki has labelled 'Fire and Flash' has received a mixed reception. It has been voted 'Dog of the Year' by a panel of world motorcycling journalists, while owners have risen to its defence by bombarding the press with letters of praise. Who is right? It all depends on what you are looking for in a motorcycle. The EN500 is certainly no use to anyone wanting a racer replica or high speed tourer, but for a relaxed lifestyle and lashings of chrome and tinsel it can't be beaten.

Above
The 1983 Z750LTD was very much a rehash of the dohc roadster twin which Kawasaki tried to sell in the late 1970s without much success – the custom version didn't fare much better

Right
In 1990 the EN450 became the EN500, but the two models share much in common such as the liquid-cooled, 8-valve dohc twin engine and belt final drive

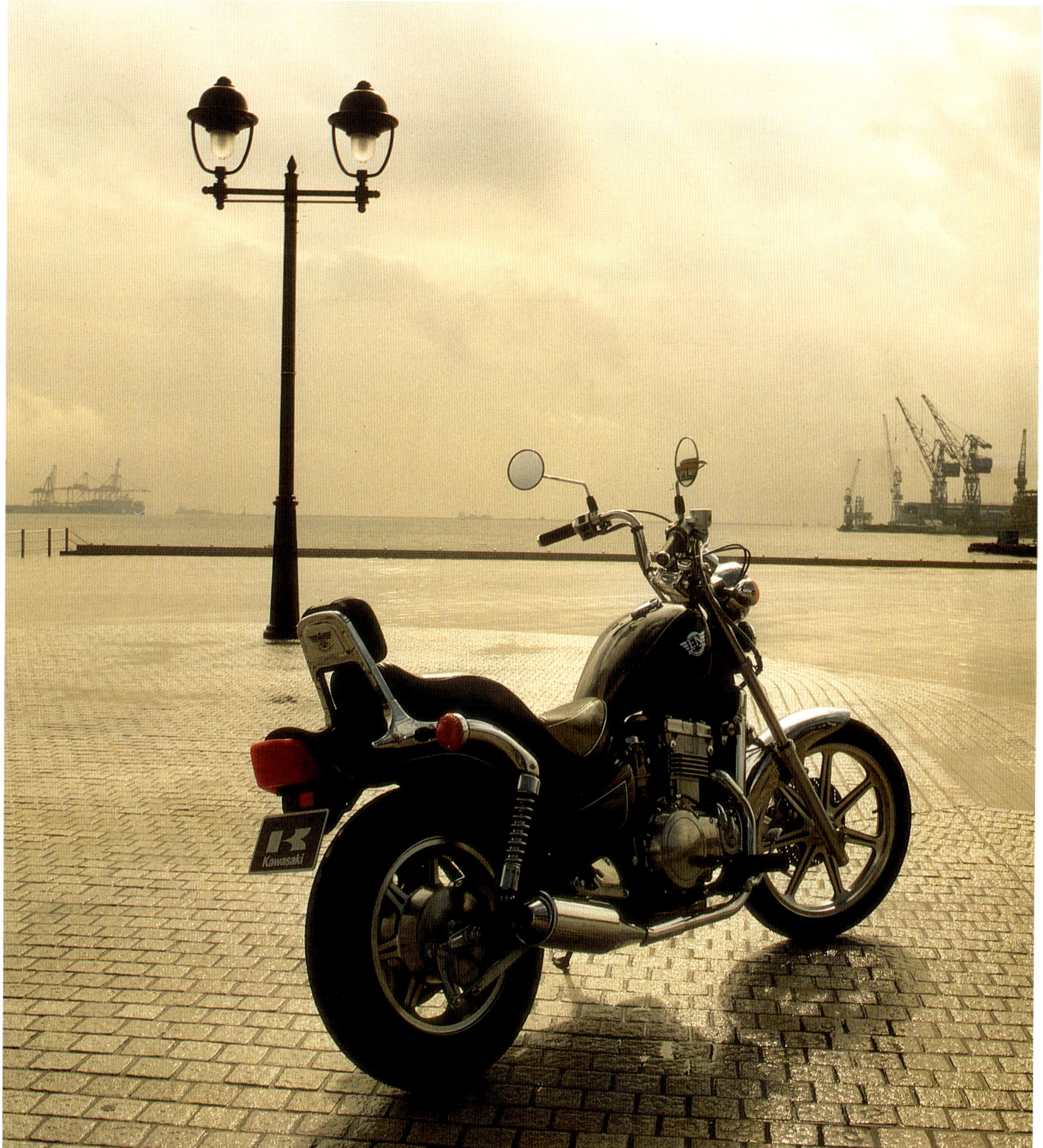
Kawasaki

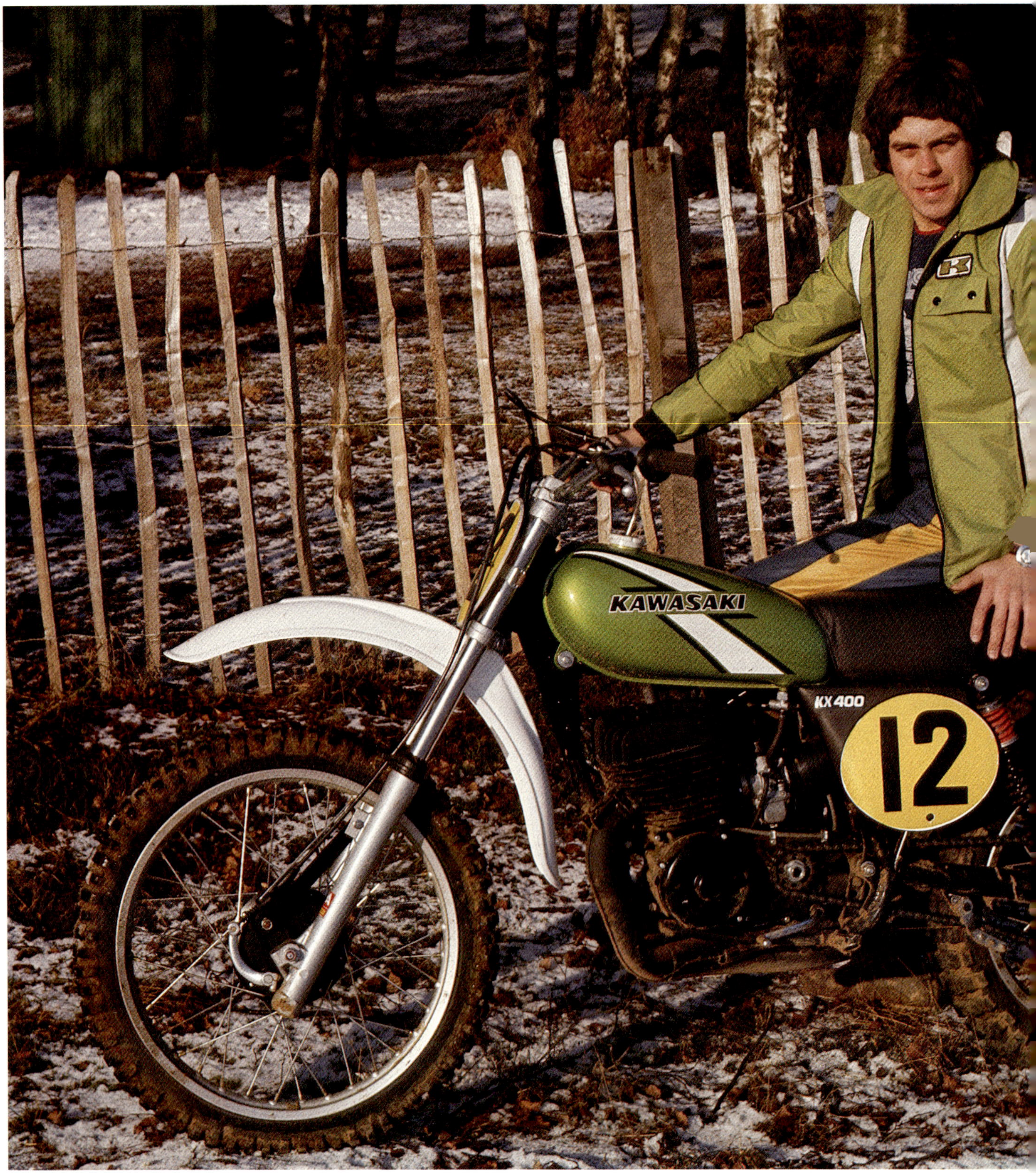
KAWASAKI
KX400
12

Off-road competition

Kawasaki's competition history can be traced back to the Meguro link and that company's success in the 1957 Mount Asama races. In that year Meguro gained victory in the 500 cc event with one of its overhead cam singles. With the birth of the Kawasaki motorcycle division came the need to publicise the fledgling organisation. The result was a number of specially prepared B8 125 cc roadsters which, kitted out for motocross events, succeeded in taking the 1963 Japanese national title. In fact these converted roadsters took the first six places in the 125 cc class. Even today their bright red tanks are recalled by older Japanese enthusiasts for the way they dominated the 1963 motocross season.

From this early success Kawasaki followed Honda, Suzuki and Yamaha into road racing, but like the others eventually found its way back into dirt bikes.

Into the early 1970s and Kawasaki designed and built a number of motocross prototypes, eventually good enough to contest the 250 and 500 cc World Championship series ridden by the likes of Hansen, Lackey and Petterson. Kawasaki powered machines also took part in the 750 cc sidecar motocross category using specially prepared three-cylinder motors; their best place was a 4th in the title chase.

By 1974 the off-road Kawasakis were performing well enough for Jim Weinert to win the important AMA motocross championship in the States. Fellow American Hansen also won a couple of 250 cc GPs that year.

By this time the company had also entered the trials world with the likes of British champion Don Smith and Nigel Birkett; in 1975 Colin Dommett took the British Experts sidecar title with his Kawasaki powered outfit. The company also offered a production version of its trials bike for sale, the KT250, over a five year period in the mid-late 1970s.

But it was towards the fields of motocross and enduro that Kawasaki directed its expertise in the off-road competition world, both in terms of works supported riders and over-the-counter sales.

Stars such as Brad Lackey, Georges Jobe and Kurt Nichol have all come close, but never won a world title for Team Green. The much publicised move by multi-World Champion Dave Thorpe from Honda to Kawasaki

A production KX400 motorcrosser, one of the very first 500 cc class dirt racers from Team Green, seen in the winter of 1976

gained many column inches, but little in the way of track results. However, thanks in no small part to its marketing expertise and a certain Alec Wright, Team Green has proved a hit with the paying customer. In both adult and schoolboy motocross Kawasaki has consistently outsold its Japanese rivals and gained a myriad of national and club championships along the way.

Formerly a trials sidecar rider with the British Greeves concern in the 1960s, Alec Wright joined Kawasaki's British arm in the following decade. By its end he had already made a significant impact as Competition Department Manager.

Take the 1979 season; in the Schoolboy field, Kawasaki won no less than nine National Championship awards with 125 cc machines, including the three main Expert Championships – the British 2-Day, ACU and BSMA National events. Wright had also been instrumental in grooming future stars such as Dave Thorpe and Jem Whatley. Led by Alec Wright, Team Green progressed throughout the 1980s, helped by innovations such as the Motocross Rider Training Scheme, Mechanic Schools and Competition Dealer Teams.

On the enduro front the company has not seen fit to offer rider support as it has done in motocross. Even so, there has been a succession of machinery on offer to the buying public, including the KDX420, followed by the KDX175, 250 and air-cooled KDX200. More latterly a liquid-cooled version of the long running KDX200 and most recently the new KDX250 (also liquid-cooled). Along the way there was also the four-stroke KLX250 of the early 1980s. Of all these the most successful in terms of sales at least, has been the KDX200, which has established itself as the 'Clubman's Choice'.

It would be true to sum up Team Green's efforts as being market leaders in the dirt bike sales league, even though, to date, they have never won a World Championship title. Instead of success by works stars on machinery which is not for sale to lesser mortals, Team Green has offered the buying national and club rider competitive, reliable and affordable machinery on a consistent basis, backed up by possibly the finest after sales service in the industry.

Above
American Brad Lackey being interviewed at the 1979 500 cc British Grand Prix

Above right
Prototype air-cooled KX500 raced by Dave Thorpe, late 1979

Right
Britain's Team Green youth co-ordinator Stuart Nunn with a line-up of schoolboy motocross talent

KX
Kawasaki

TEAM GREEN
M.ROBERTS
DUNLOP
Racing
RENTHAL
SCOTT
RIZLA+
NGK
DUNLOP

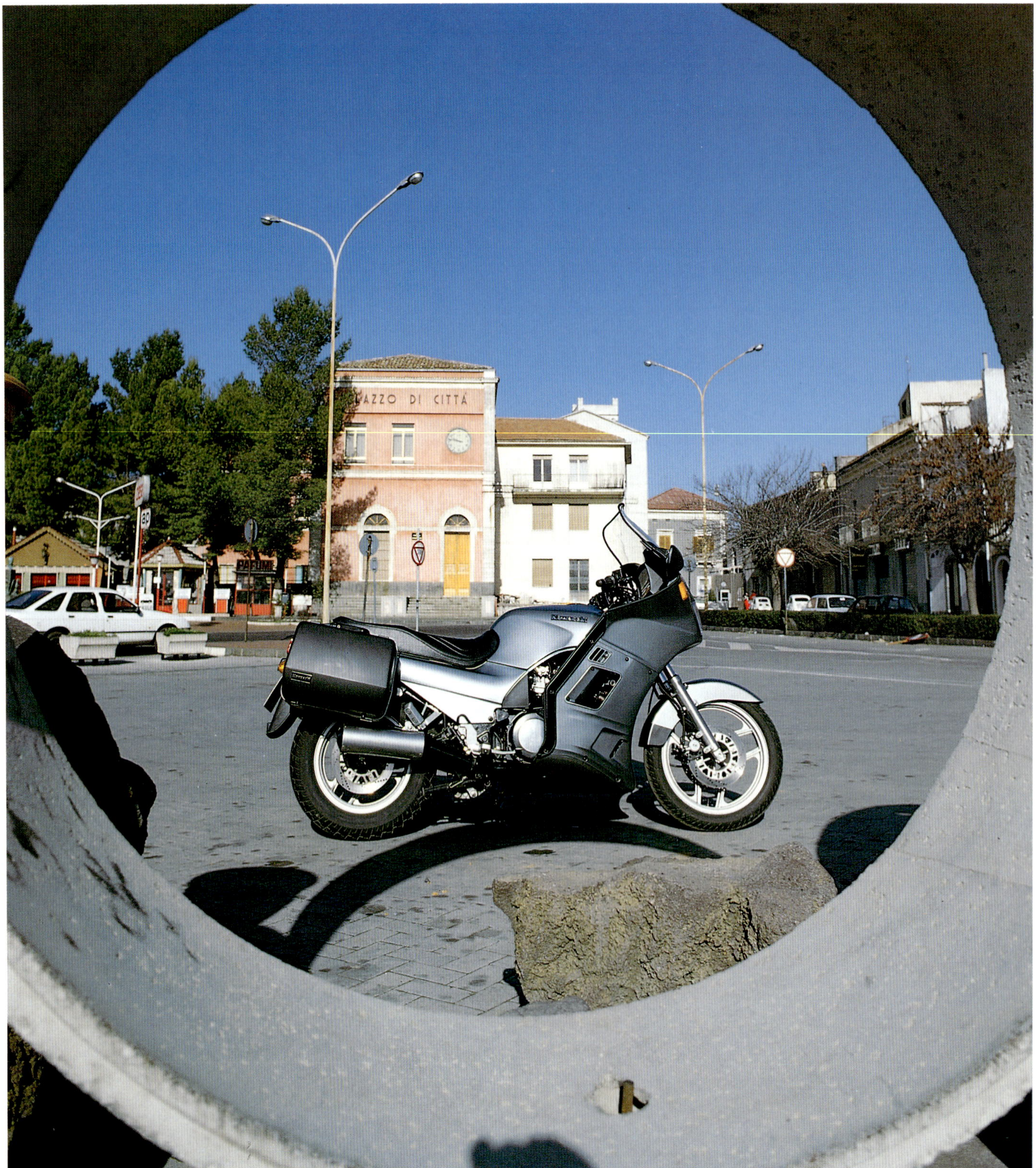
AZZO DI CITTÀ

GTR

Launched in the spring of 1986, the 1000GTR (coded ZG1000-A1) was designed for the rider who wanted to travel fast without sacrificing the comfort, convenience and load-carrying capacity of a touring bike.

Essentially the GTR is a combination of a detuned GPZ1000RX powerplant, new diamond frame, shaft drive and radial tyres.

The 997 cc (74 × 58 mm) dohc GPZ1000RX-based engine runs on a compression ratio of 10.2:1. Maximum power of the original A1 version was 108.5 bhp at 9000 rpm, with maximum torque being produced at 6500 rpm. Kawasaki claimed 135 mph and the standing 400 metres (quarter mile) in 12 seconds. In Germany and Sweden the power was restricted to 100 bhp and in Switzerland 70 bhp. An oil cooler is standard in all markets.

To minimise torque reaction during high speed work, the shaft/swinging arm is exceptionally long – 528 mm (20.8 in) from pivot centre to axle centre.

Similar to the frame used on the GPZ900R, the GTR's diamond frame utilises the engine as a stressed member of the chassis. In this design, the downtubes are eliminated and the engine is mounted lower, cutting

Left
Easy on the eye and eminently practical, the 1000GTR was Kawasaki's first purpose-built touring bike

Right
The GTR is actually a better bike two-up than solo

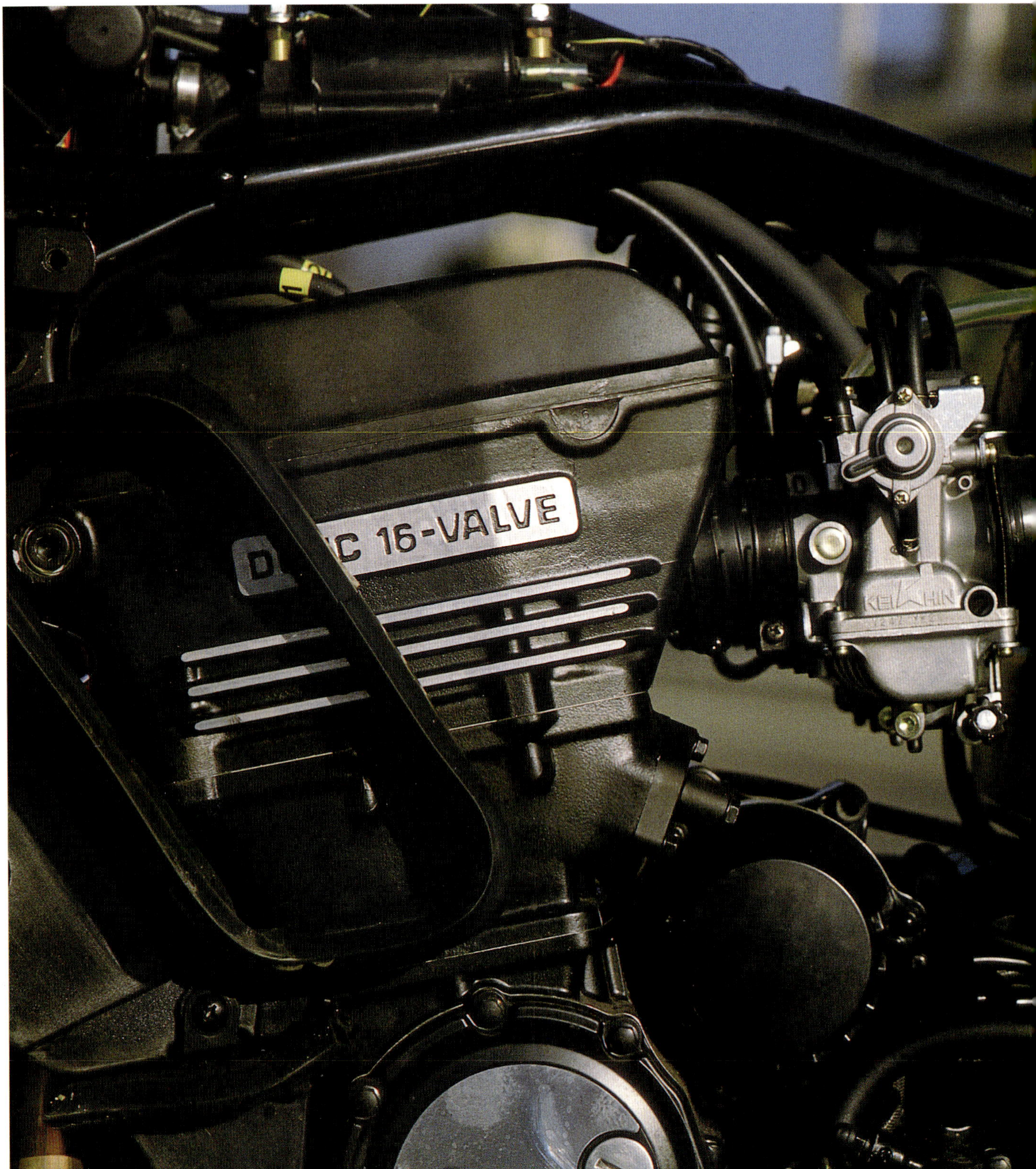
16-VALVE

weight and lowering the centre of gravity in an attempt to improve handling – which from the author's experience is still judged on the heavy side.

The one-piece main frame is constructed of high-tensile steel, and consists of steering head, backbone tubes, swinging arm pivot, and rear section support. A short bolt-on alloy rear section supports the dual seat and panniers.

The Dunlop radial tyres were the first of their type to be fitted to a standard production Kawasaki, offering improved stability at high speed over conventional tyres of the era.

Fitted as standard equipment the large fairing provides a fair level of protection against the elements. It boasts an extra large screen, flush-mounted headlight, integrated direction indicators, digital clock and two storage compartments.

A large 28.5 litre fuel tank provides sufficient range for the serious touring rider, particularly when a sedate riding stance is adopted.

Each pannier holds up to 10 kg (22 lb) and can take a full-face helmet. The panniers are removable in seconds by simply flipping a single latch, and integral handles make them easy to carry when not on the bike. When the panniers are removed, special body panels are provided to cover the mounting points. There is what Kawasaki describe as a mini-carrier which has a maximum capacity of 5 kg (11 lb).

Air-adjustable Uni-Trak rear suspension features 4-way rebound damping, while at the front the air-assisted forks have 41 mm stanchions, plus an equaliser for easier adjustment.

The 12-volt electrical system comprises a 400-watt alternator, electric starter, electronic ignition, 60/55-watt halogen headlamp and 18-amp-hour battery.

Braking is taken care of by a pair of 270 mm discs up front and single 280 mm at the rear, all with sintered metal pads. The six-spoke aluminium wheels are 18-inch at the front and 16-inch rear.

A number of accessories are available. These include a Tote bag which can be strapped to the rear carrier or saddle, a two-part Tank bag (which when used together provide 20-litres of carrying capacity), and Pannier bags (designed to fit inside the pannier cases); the latter come as standard equipment on the GTR.

The GPZ1000RX-based engine runs on a compression ratio of 10.2:1. Maximum power of the original A1 version was 108.5 bhp at 9000 rpm, with maximum torque being produced at 6500 rpm

Colours for the A1 model were Pearl Gentry grey or Candy Wine red.

For 1988 the GTR received the A2 prefix; the following year saw the A3 and in 1990 it became the A4. In technical terms the GTR has been virtually unchanged.

What has changed are the colour schemes which have included Pearl Alpine white, Candy Persimmon red, Luminous Polaris blue and even for a short period Dark green. The other virtually un-noticed change has been to fit an ever increasingly restrictive exhaust system, which has seen the power output of the unrestricted market version drop to a low of 84.5 bhp for 1992.

Even so, the GTR remains a popular choice with touring riders offering a viable (and considerably cheaper!) alternative to BMW's K100LT model.

It's owners see it as very much a serious touring machine, designed to be equally adept at cruising down the autostrada with a pillion rider and a stack of holiday luggage as it is threading through traffic jams on the way to work.

It also looks likely to fulfil its makers original '10 year lifespan' target; certainly the lack of changes bodes well for existing and new owners alike.

What owning a GTR is all about – the open road and sunny climes

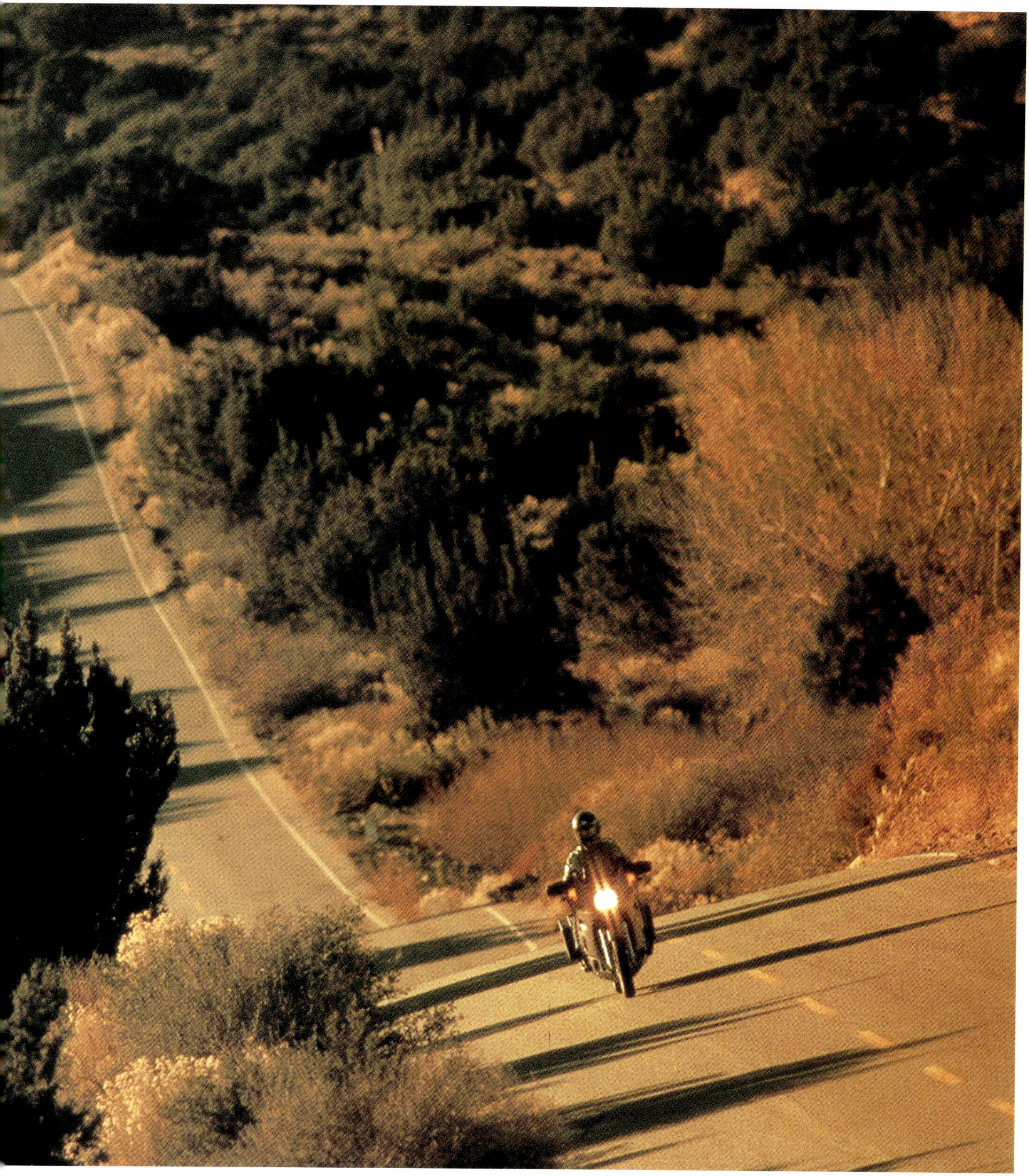

Above
In many ways the GTR's nearest rival is BMW's K1000RT. Kawasaki is cheaper and faster; the German bike is more expensive and includes 'snob' value

Right
The large fairing provides the rider with a fair degree of protection, less so for the pillion passenger

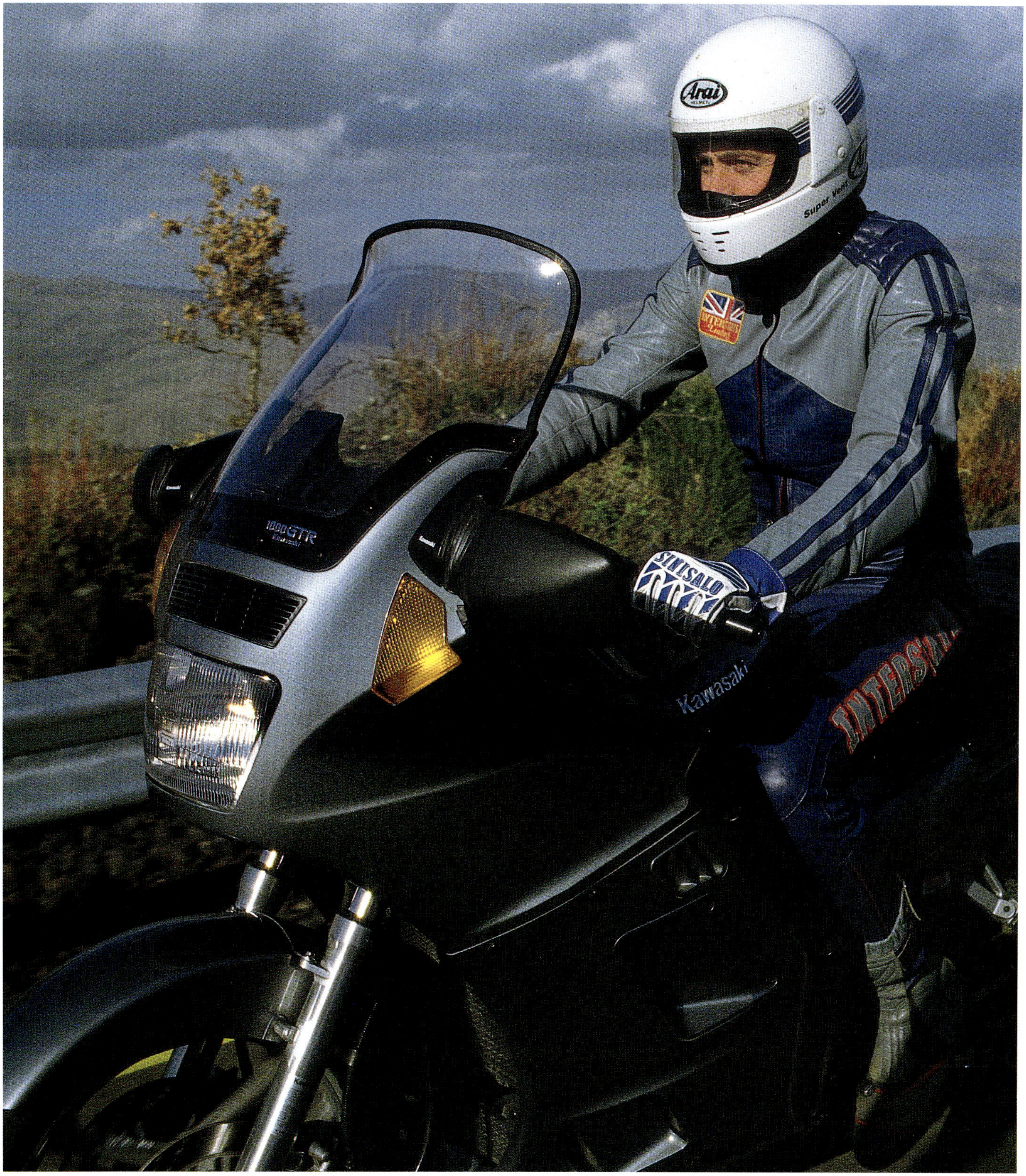
Arai
Super Vent
1000GTR
Kawasaki
SINISALO
Kawasaki

Liquid-cooled fours

If the legendary Z1 was the father of the air-cooled four-cylinder Kawasaki family, then the GPZ900R is without doubt the sire of the company's liquid-cooled models. In fact, its hard to believe that this latter machine was shown to the Press for the first time in December 1983, with production versions going on sale early in 1984 – the GPZ900R seems to have been around much longer, such has been its influence on modern Superbike design.

Known as the Ninja in the USA and Canada, the 900R was the first Superbike with a liquid-cooled, 16-valve dohc in-line four-cylinder engine. It was also the first to combine a lightweight diamond frame, aluminium rear frame section and 16-inch front wheel. And, Kawasaki claimed, the first with a fork that delivered truly progressive wheel travel combined with their rising rate Uni-Trak rear suspension.

Lighter than some 750s, more powerful than some 1100s, it was the fastest production Kawasaki streetbike at the time of its launch.

Continual development of the various big-bore models using the basic air-cooled 8-valve across-the-frame four-cylinder engine mounted in a conventional double-cradle frame had kept Kawasaki at the top of the Superbike stakes ever since the Z1. However, realising that time and technology would eventually catch up, the company's engineers began a six-year long programme to develop a suitable replacement. Although several other configurations – including a V4, V6 and an across-the-frame six – showed promise in certain areas, all were ultimately judged to offer no significant advantage over the across-the-frame four.

So the proven configuration was retained, but virtually every facet of the layout re-evaluated. The result was an all-new engine which although sharing the inherent advantages of the conventional four (including relatively few moving parts, low mechanical loss, efficient combustion flow, evenly-spaced power strokes, excellent primary balance and relatively straight forward servicing), redressed a major drawback of previous designs: that of width.

Several design innovations meant that Kawasaki engineers were able to create a 908 cc motor producing almost 114 bhp and 8.7 kg/m (62.93 ft/lb) of torque, with a width of just 541 mm (17.76 in). This was a full 123 mm

A GPZ900R in the paddock at the 1984 Isle of Man TT. Dealer-entered 900Rs took the first three places in the Production race that year

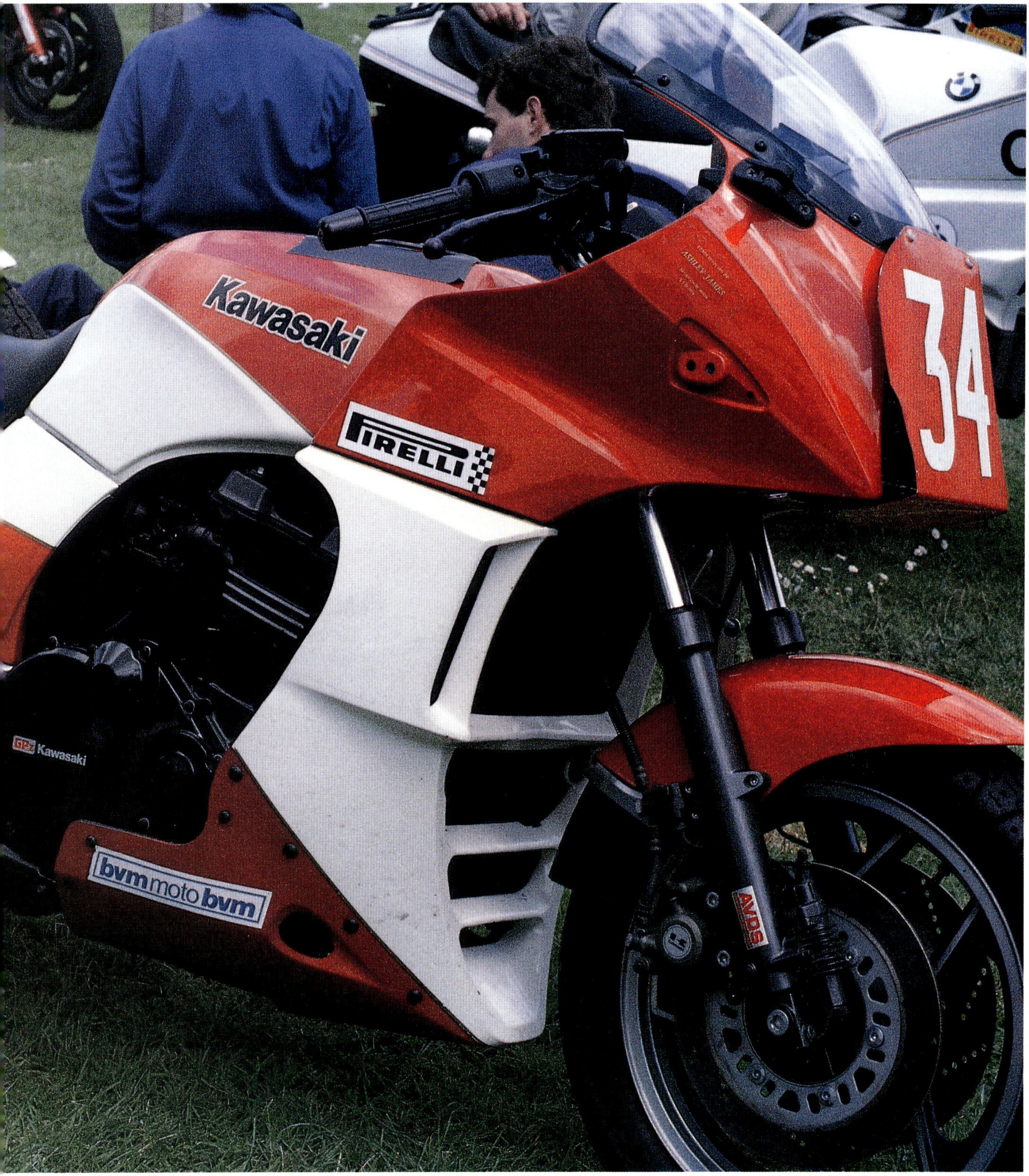
Kawasaki
PIRELLI
34
bvm moto bvm
GPz Kawasaki
AVDS

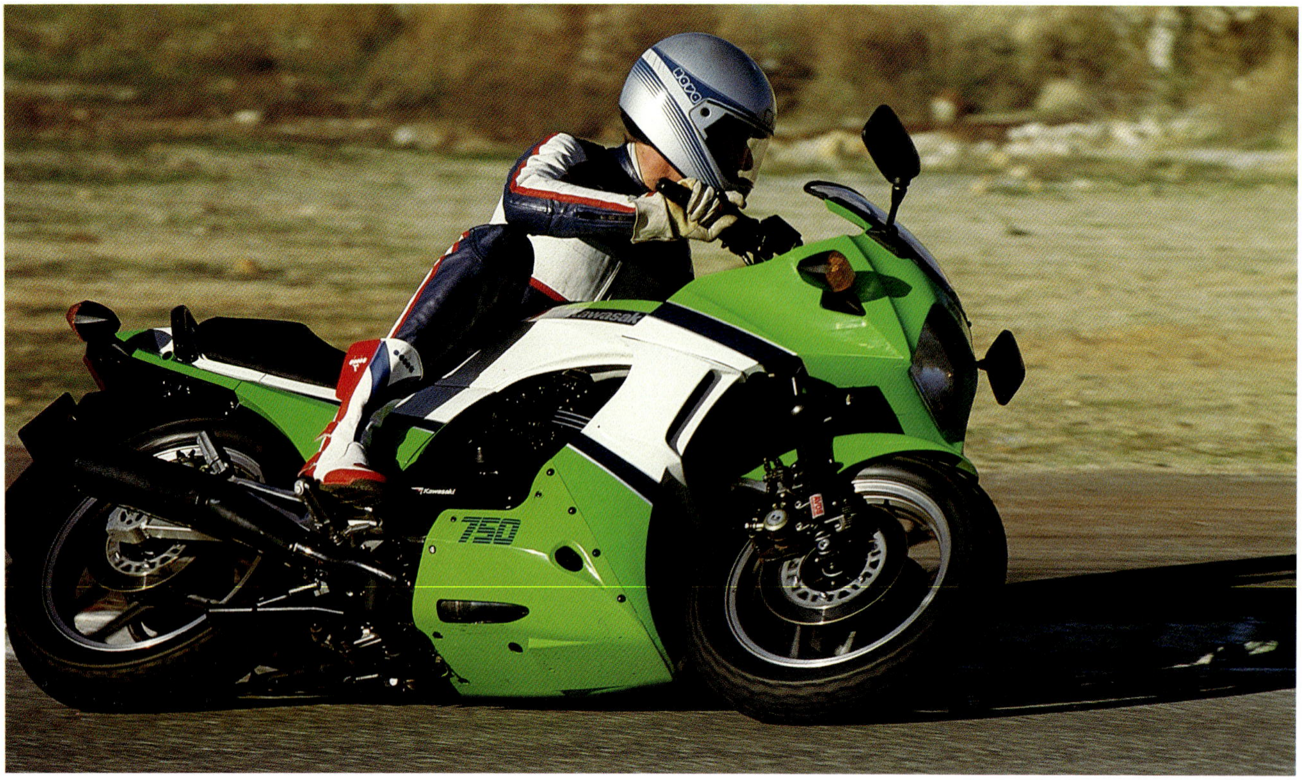
750

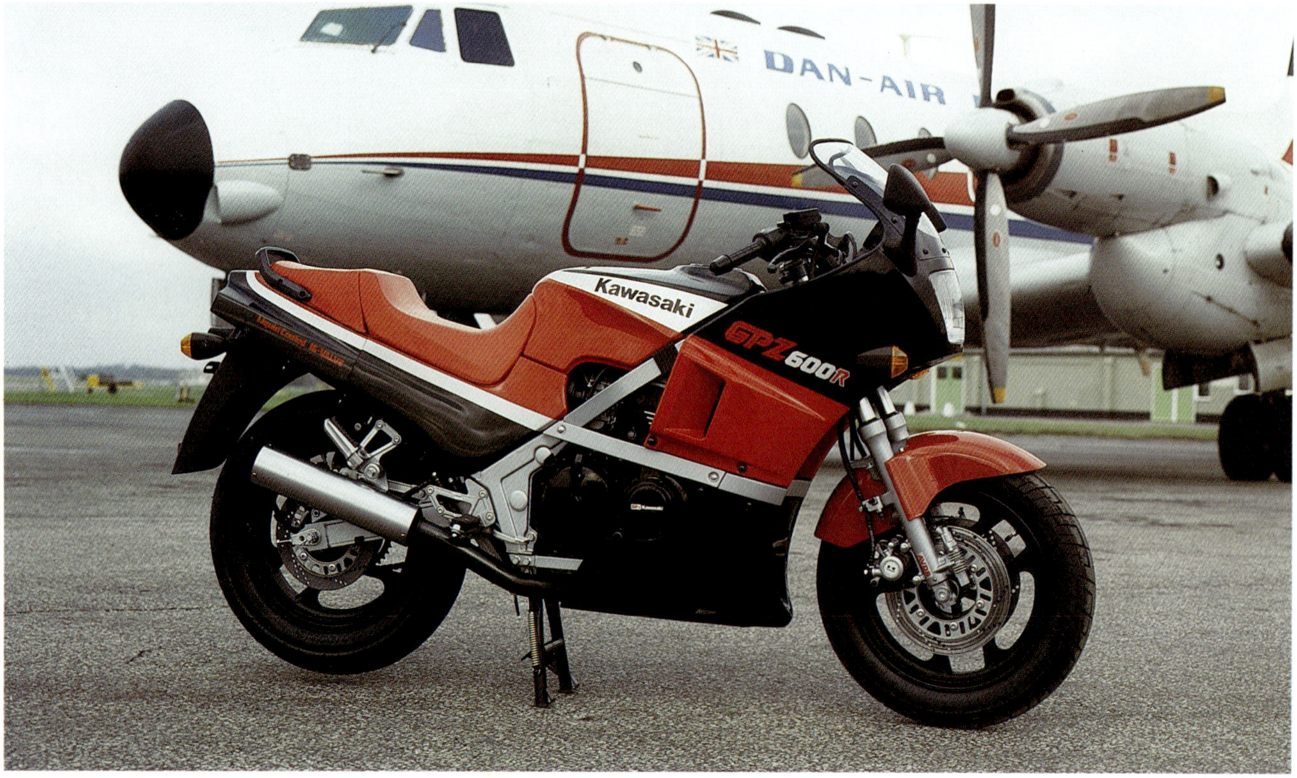
DAN-AIR
Kawasaki
GPZ600R

Above

An American journalist testing one of the first GPZ900R's (Ninja in the USA) to reach the States at Leguna Seca in early 1984. The bike set new standards in both performance and handling

Above left

Following the success of the 900R, Kawasaki introduced the GPZ750R. Except for its smaller capacity it was essentially the same bike; except in Japan, the 750R never sold in large numbers

Left

Next in the liquid-cooled GPZ family came the 600R, which shared numerous features with its bigger brothers, such as dohc and 4-valves per cylinder. However, it was far from being simply a scaled-down 750/900R, but very much a new machine in its own right

(4.86 in) narrower than the original 903 cc Z1 powerplant, despite oversquare bore and stroke dimensions of 72.5×55 mm. This in turn meant that the new engine could be mounted about 30.5 mm (1.2 in) lower, for quicker handling, without sacrificing cornering clearance.

Although the included valve angle had been decreased more than 25 degrees (compared to both the 1983 GPZ750 and GPZ1100 air-cooled models) the GPZ900R engine was shorter top to bottom than the Z1. The new engine was also more compact, and including aluminium radiator, oil cooler and other engine components, was some 5 kg (11 lb) lighter than the Z1.

A major factor in this compact layout was a basic change to the across-the-frame four layout. Liquid-cooling (a mixture of water and anti-freeze) meant that the cam chain could be repositioned outside the cylinder bank, giving the 900R/Ninja a wet-liner cooling system which was both compact and efficient. Another space-efficient feature was the air-cooled alternator, mounted above the gearbox and driven by a simplex chain off the right side of the crankshaft.

For maximum reliability up to the 10,500 rpm red line, the one-piece crank turned in five plain bearings, with inserts made of the same alloy as the larger GPZ air-cooled models. In addition to driving the clutch, the crank's primary gear also drove a compact counter-balancer (the first on any across-the-frame four) which smoothed out secondary vibration. The teeth on both the primary and balancer gears were polished for reduced mechanical loss and minimal backlash.

Liquid-cooling allowed both decreases in piston clearance and ring tension and a power-boosting 11:1 compression ratio.

Also with five bearings each, the hollow, lightweight camshafts turned in the head itself. Each camshaft had four lobes, actuating two valves each through dual-finger cam followers. Valve adjusters were of the solid screw-and-lock nut variety.

Kawasaki claimed the new constant-load automatic cam chain tensioner ensured decreased mechanical loss and more constant chain tension, while the 7.94 mm chain pitch gave more strength than a conventional chain, but without the extra weight.

Another advantage of positioning the cam chain outside the cylinder bank was that the four combustion paths from airbox to exhaust pipe are as short and straight as possible. This also made possible symmetrical sub-ports for each set of inlet and exhaust valves, further improving breathing efficiency. Another boost in this area was provided by the all-new compact Keihan CVK 34 mm carburettors with their semi-flat slides ensuring improved fuel atomization, quicker throttle response and reduced intake resistance. A full 19 mm (.75 in) shorter than conventional 34 mm carbs, these aluminium-bodied units also cut weight.

The liquid-cooling system utilised a single aluminium cross-flow radiator, thermostat, electric fan and water pump. The latter component was gear driven off the clutch from the same shaft which operated the oil

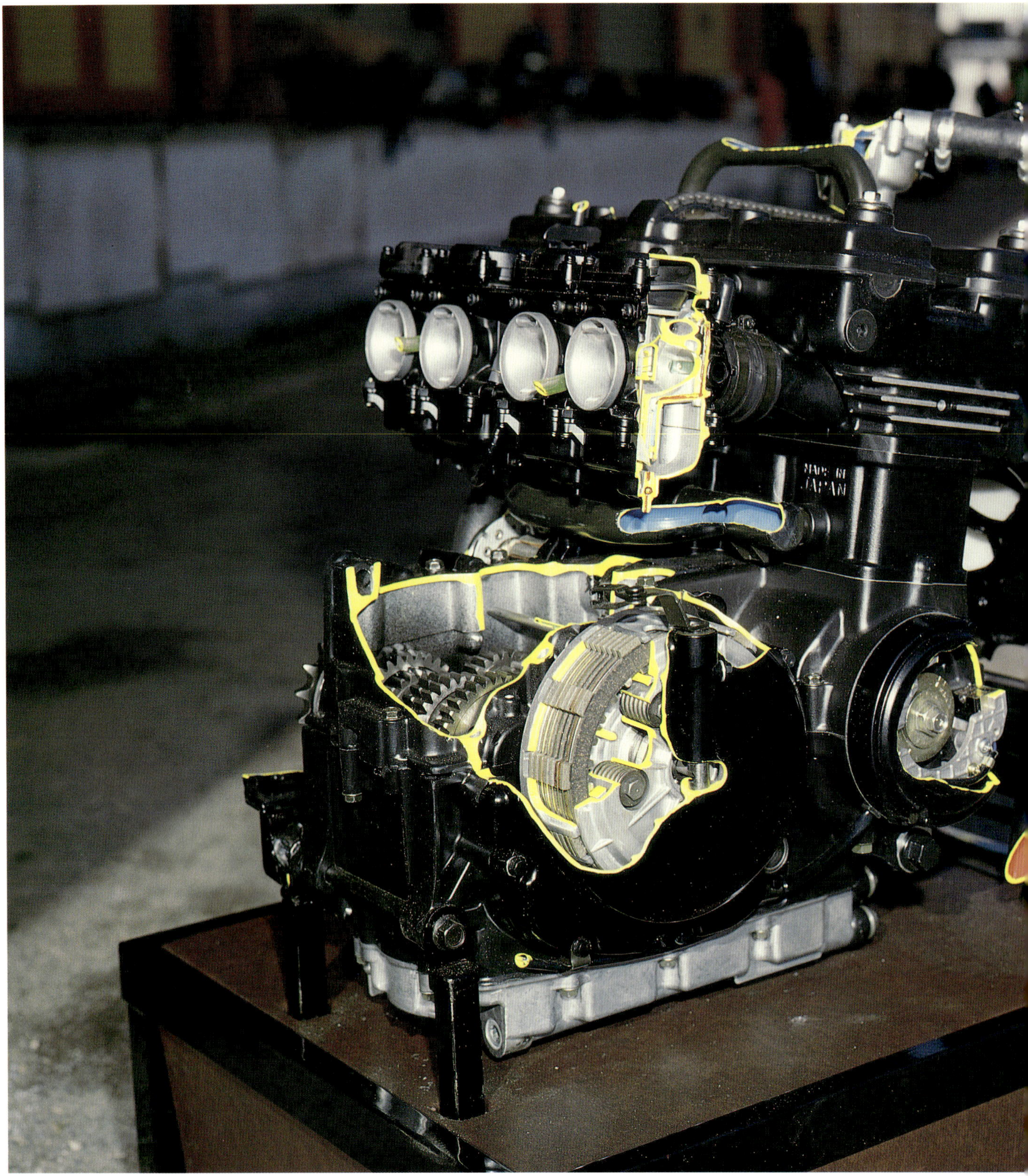
MADE IN
JAPAN

Above
Unlike the 750 and 900, the motor in the 600R simply 'sat' in its chassis, and used rubber engine mounts at the front – as per the air-cooled 550. However, the 600R chassis was no throw-back to an earlier era. Instead, it was a double-triangulated perimeter frame of steel construction and mostly rectangular in section

Left
The 592 cc (60 × 52.4 mm) engine, although having many similarities with the 750/900R, was actually based more closely on the air-cooled GPZ550's powerplant

pump. Lubrication was by a special dual-stage system. In the primary oil loop, oil was drawn through the primary filter to the pump, then passed through the filter element to the crank, gearbox, head and other major components. In the secondary loop, oil is passed from the pump to the four-tier oil cooler, then back to the sump.

The oil cooler loop serves a secondary function as a temporary 'storage tank'. Sump oil level is kept at a minimum, while a special dam isolates the critical area around the balancer gear, primary gear, and clutch, a cunning solution.

As well as a new hydraulically operated clutch, the 900R/Ninja featured

107

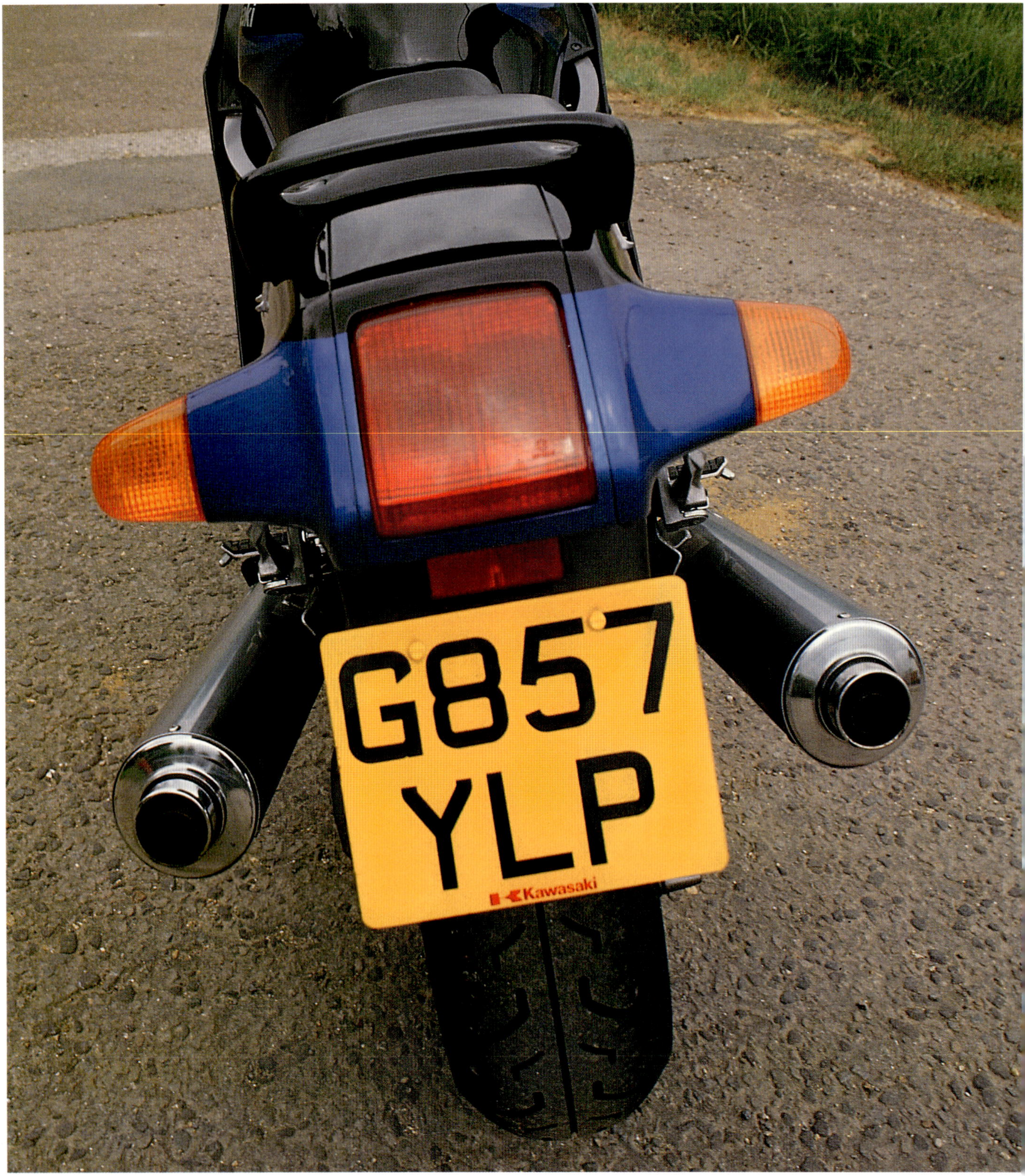
G857
YLP
Kawasaki

Left

A reminder of where all this development is leading . . . Kawasaki, like all good companies, realised that liquidity is a good thing, and in their case it would lead to the fastest production motorcycle in the world. This is the machine ahead of them: the superb ZZ-R1100

Right

The 1987 Kawasaki UK entered team pose with their GPX750s and back-up crews during the Isle of Man TT

a 6-speed gearbox, another first for Kawasaki on a big-bore machine. The gearbox shafts are staggered in the vertical plane to save space and reduce mechanical loss.

The diamond frame was constructed in high-tension steel tubing, which Kawasaki claimed, was stronger, yet lighter, than conventional steel tubing. There was a backbone sub-frame that stretched over the engine from the steering head to swinging arm. The engine is mounted to this backbone at six points (four rear, two front).

During the early stage of frame testing, conventional downtubes were added to determine what stress, if any, they would be subjected to. These results showed the downtubes carrying virtually no load and both were removed from subsequent testing.

Other technical details include: Automatic Variable Damping System (AVDS) anti dive for the front forks, a fork stanchion of 38 mm diameter (1 mm up from the air-cooled 750/1100 GPZ models), aluminium box-section swinging arm, six-spoke alloy wheels, V-rated tyres, triple disc brakes (a pair of 280 mm diameter discs at the front, a single 270 mm component at the rear), 22-litre fuel tank and a dry weight of 228 kg (503 lb).

The press and public alike gave the newcomer full approval, thanks in no small part to its excellent performance (*Motor Cycle News* got 158 mph) and sales got a further boost when in June 1984 (just three months after its release on the British market) further proof came of the 900R's sporting ability when three dealer entered models were first past the post

Above
Striking red and white paint job of the GPX750's angular fairing

Right
For 1988 the GPZ1000RX was upstaged by the even more powerful ZX-10. With 136 bhp, increased torque and higher maximum speed, the ZX-10 took over the mantle as the 'world's fastest production roadster'

in the Production TT in the Isle of Man. Headed by race winner Geoff Johnson, the trio of 900Rs finished ahead of all the works supported teams from rival manufacturers.

To round off a truly memorable year, the readers of *Motor Cycle News* (the world's top selling motorcycle newspaper) voted the 900R as their 'Machine of the Year'.

But the 900R wasn't perfect: early examples suffered a spate of warranty claims caused by a faulty camshaft case-hardening and associated problems.

Next in the liquid-cooled GPZ family came the 600R, which shared numerous features with its bigger brother – such as dohc and 4 valves per cylinder. Far from being simply a scaled down 900R, it was very much a new machine in its own right.

The 592 cc (60 × 52.4 mm) engine, although having many similarities

110

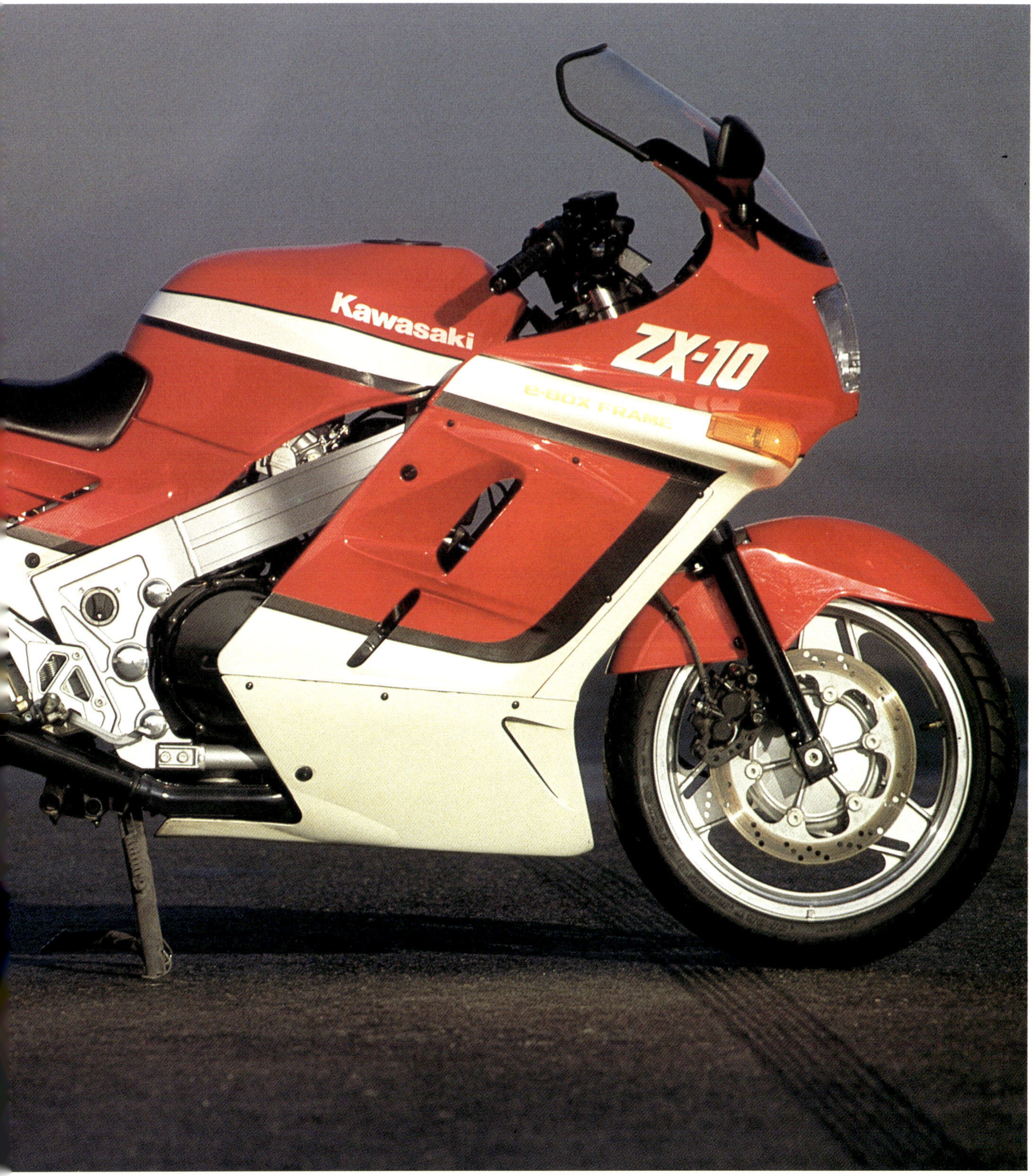
Kawasaki
ZX-10
E-BOX FRAME

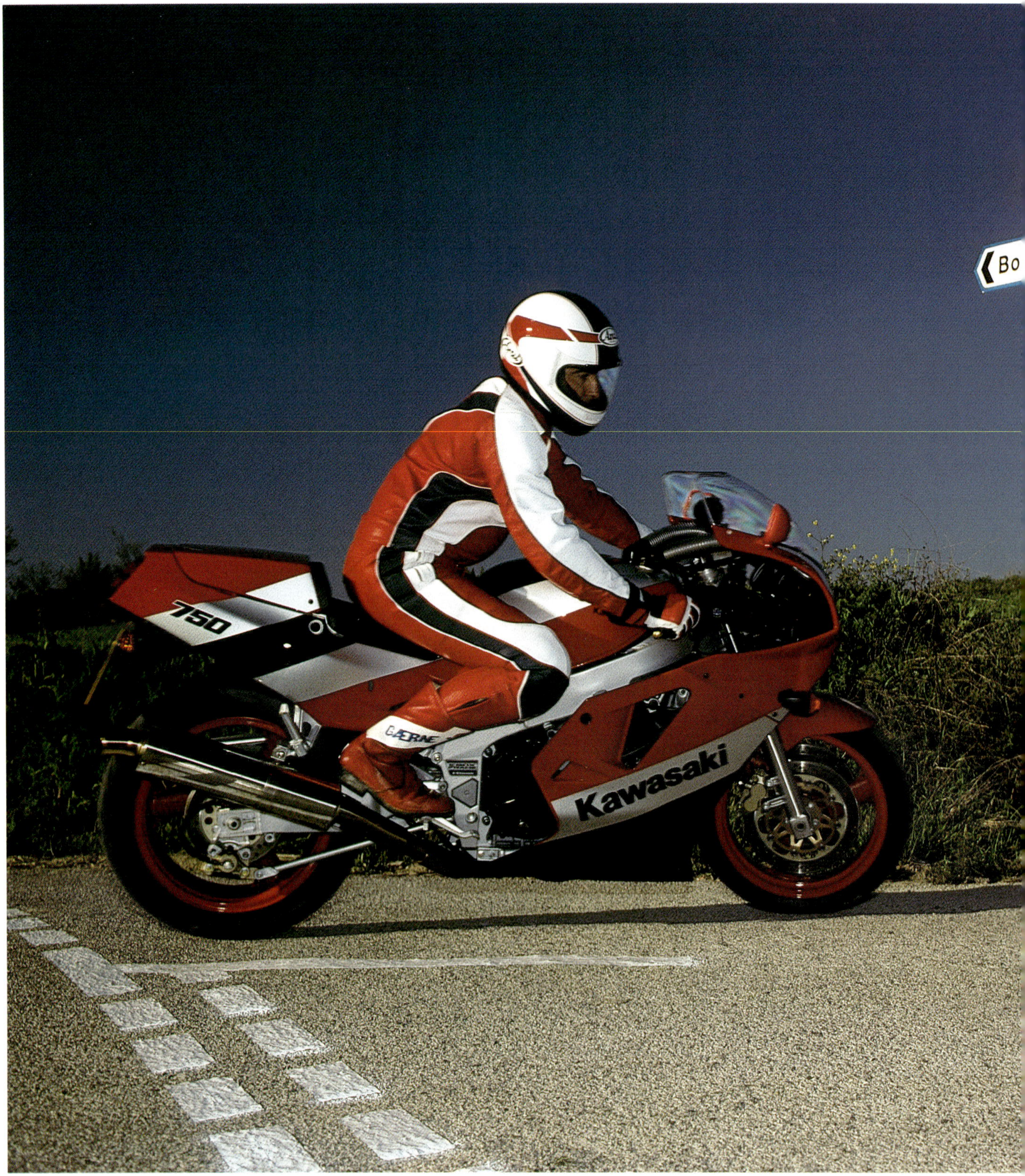
750
Kawasaki

with the 900R, was actually more closely based on the air-cooled GPZ550's powerplant. The liquid-cooled 600R developed 75 bhp, ten more than its air-cooled brother the 550, while torque was also considerably increased, albeit mainly above 7500 rpm.

How had this been achieved? Bottom end of the 600 was virtually the same as the 500, but the use of a new and slimmer alternator, allowed overall width to decrease by 40 mm (1.5748 in).

The top end benefited from a number of changes, many of these bringing it into line with the 900R (which had by now been joined by a smaller 750R), though the 550's centre cam chain location was retained.

Like the 750/900R, tappet clearances were made by screw adjusters, the carbs had aluminium bodies and semi-flat slides; but a major difference between the two engines was that the 600 did not have a balance shaft. Why? Because the chassis of the two machines were different. The 750/900R had the diamond type that used the engine as a stressed member, and so vibration-absorbing rubber engine mounts could not be used. In contrast, the 600R motor simply 'sat' in its chassis, and used rubber engine mounts at the front – as per the air-cooled 550.

However, the 600R chassis was no throw-back to an earlier era. Instead, it was a double-triangulated perimeter frame, of high-tensile steel, mostly rectangular in section.

Using computer-aided design technology, Kawasaki engineers wrapped the powerplant with four main tubes, plus strategically placed cross braces, to keep the steering head and swinging arm pivot in optimal alignment. The distance between these two critical points was 50 mm (1.9685 in) shorter compared to the 550, helping to reduce frontal area - as did the slippery aerodynamics of the fairing and the use of 16-inch wheels.

Like the 900/750R, the 600 also featured the AVDS fork and Uni-Trak rear suspension. Other details included V-rated tubeless tyres, triple disc brakes, 18-litre fuel tank and a dry weight of 195 kg (430 lb). Maximum speed was a whisker under 140mph.

In 1987 the GPZ600 was joined by the GPX600. Modifications to the engine resulted in it not only being 1.5 kg lighter, but also 13 per cent more powerful. Improved design to the inlet porting, new inlet valves, lighter pistons, chrome-molybdenum connecting rods, and a larger airbox were some of the major improvements. To match the increase in horsepower, the radiator's cooling capacity was upped by 23 per cent.

Utilising technology developed by the factory race team in the 1988 World Endurance Championship, Kawasaki introduced the new ZXR750 the following year

Maximum power, 84 bhp at 11,000 rpm, top speed 144 mph. There was also a new frame, new fairing, new front suspension, new disc brakes and new cosmetics/styling; much of this being based on the GPX750 which had been introduced at the end of 1986. The GPZ continued in production and was uprated in the engine department to virtually GPX standard for the 1989 model year. Both were discontinued at the end of that year to make way for the new ZZ-R600.

Having already created liquid-cooled fours for the 600, 750 and 900 classes it was perhaps obvious that Kawasaki would expand their range to include the 1-litre Superbike, and so for the 1986 model year the new GPZ1000RX took over from the air-cooled GPZ1000.

The 997 cc (74 × 58 mm) engine followed similar lines to the 900R. Kawasaki claimed it to be 'absolutely the world's fastest production streetbike'.

Compared to the 900R the 1000 RX's major new features included the following:

Flat aluminium 10.2:1 pistons.
New valves, increased air flow and reduced weight.
Exclusive new air induction system drawing in only the coolest, cleanest air possible.
Hand polished inlet ports.
Four larger Keihan CVK 36 mm carbs.
Stainless steel exhaust pipes.
New mild steel silencers with corrosion-resistant dull-chrome finish.
Five-tier oil cooler.
Two rubberised engine mounts and compact balancer to reduce vibration.
Stronger clutch.
Undercut dogs on third through to sixth gears for most positive shifts.
632 O-ring chain.
New perimeter frame.
New fairing with drag coefficient claimed to be as low as a GP race bike.
Flush-mounted 60/55-watt halogen headlamp.
Aerodynamic integrated indicators.
Fairing ducts to direct cool air to radiator exhaust pipes and riders legs.
Front fender with 'wing' to help direct air-flow to radiator.
Modified Uni-Trak featured linkage which compressed the shock from both ends.
Swinging arm strongest fitted on a production Kawasaki up to that time.
40 mm front fork with integral aluminium brace, AVDS, air-assist and 3-way adjustable anti-dive.
Triple disc brakes (280 mm front, 260 mm rear) with improved, fade-resistant sintered metal pads.

Above right
The biggest changes of the ZXR750 compared to the GPX750 came in the cylinder head design, valve train, induction, exhaust, clutch and, perhaps most of all, the all-aluminium perimeter-type frame, similar in configuration to that of the ZX-10

Right
A feature of the ZXR750 are these massive pipes, which act as cool-air ducts to the large airbox underneath the fuel tank; a similar set-up is also found on the ZXR400

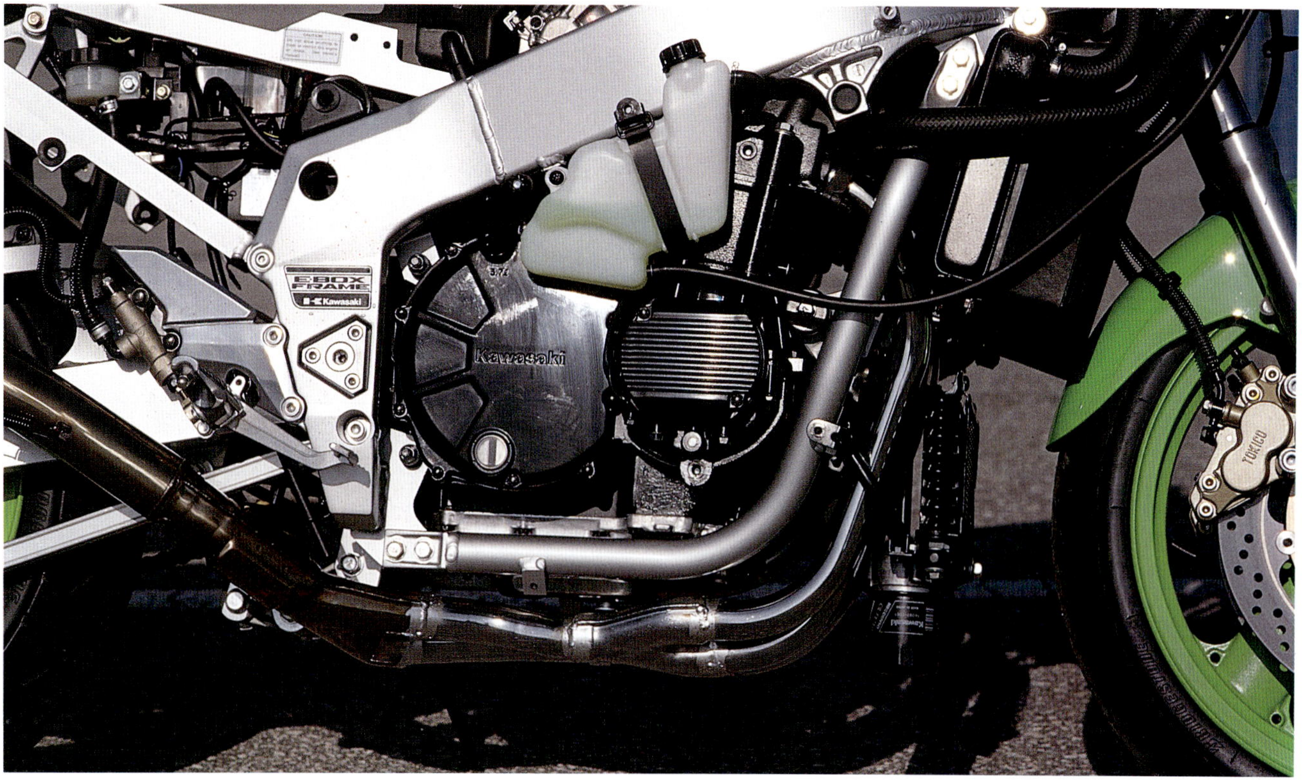
E80X
FRAME
Kawasaki
Kawasaki
TOKICO

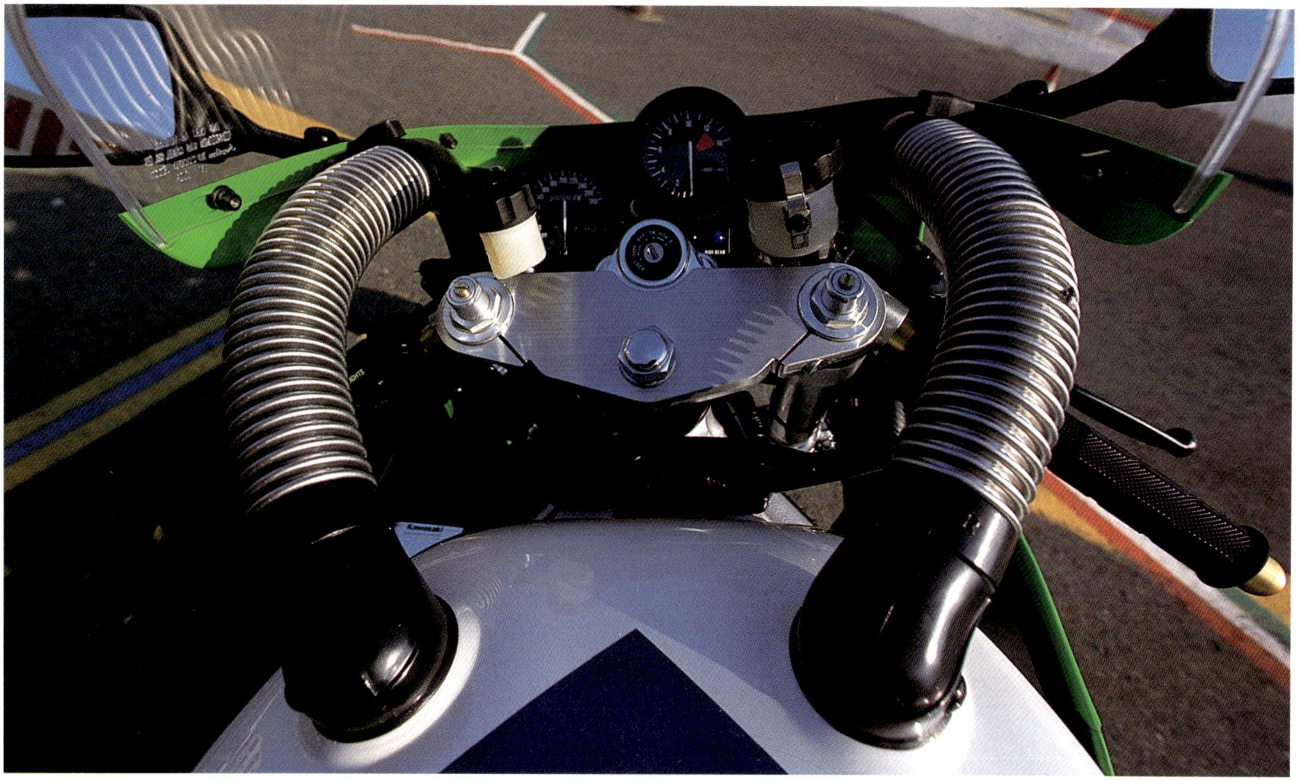

New synthetic fibre brake lines claimed to reduce expansion 20 per cent compared to standard components.
16-inch V-rated tubeless tyres (front: 120/80, rear: 150/80).
Retractable luggage hooks.
Dual helmet locks.
Retractable passenger grab rail.
Maximum power of 124 bhp was produced at 9500 rpm, the standing 400 metre (quarter mile) distance could be covered in 10.6 seconds and the top speed was 163 mph. All class-leading stuff, but out on the road, and even more so the track, the new 1000RX found itself unable to match the slower and lighter 900R from A to B. In other words it was difficult and demanding to ride in anything other than a straight line. I must admit however, that a brief ride I had on an early example left a lasting impression of its awesome power, which remained unmatched until 1992 when the author became the proud owner of a full power version of the ZZ-R1100.

Next in the liquid-cooled four-cylinder family came the 748 cc (68 × 51.5 mm) GPX750R, launched in the autumn of 1986 for the 1987 model year. With a dry weight of only 195 kg (430 lb) it was little bigger dimensionally than the GPZ/GPX600 models. Its power-to-weight ratio was also good with 105 bhp at 10,500 rpm on tap, providing a good level of performance.

There were a number of new engineering features which are explained below.

RPMS *(Redline Plus Maximum Mass-reduction System)*.
In the GPX750R's valve train, two short, hollow camshafts actuated 16 valves through individual lobes and rocker arms. Individual rockers were claimed to reduce reciprocating mass and eliminate the chance of uneven cam lobe loading. The rockers pivoted on rods, which were anchored at the sides of the head instead of the middle. This allowed the designers maximum freedom in port design and the GPX750R benefited by having smoothly curved, tapered ports. Specific RPMS benefits included greatly reduced bulk, an included valve angle of just 30 degrees, and the capacity (in theory, if not in practice) for the valve train to withstand 14,500 rpm.

HI-TECH *(High-velocity Induction Technology)*
This power-boosting induction system improved breathing efficiency in two stages. First, special ducts with patented 'ramp vents' smoothed air flow into semi-flat-slide carbs, for a non-turbulent, high-speed stream of air. Secondly, this high-velocity charge was carried to the cylinders through intake ports smoothly tapered for maximum efficiency.

Right
Factory race kit for the 1989 ZXR750H1 includes close ratio gears, generator, rear-set foot controls and lighter, less restrictive exhaust can

Below right
The beginning of a new decade saw the debut of the all-new ZZ-R series in 600 (as here) and 1100 cc categories

BACK STEP KIT
CLOSE-RATIO
TRANSMISSION
ASS'Y
Generator

SHOEI
bike
TWIN CAM 16-VALVE
ZZR
600

DADS *(Dry Alternator Drive System)*
In this new exclusive charging system, the alternator was driven by a Multi-V-belt, which was lighter than a chain and eliminated the need for a damper. DADS was also said to be easier to service and modify, and the alternator could be removed easily for racing.

FAST *(Featherweight Aluminium & Steel Technology Frame)*
As the title implies this was a combination of a high-tensile steel frame and a short alloy rear section which was actually lighter than rival manufacturers' alloy diamond frames of the time.

ESCS *(Electric Suspension Control System)*
The new 38 mm braced fork featured Kawasaki's first electric suspension system. ESCS valves opened and closed automatically according to the inner oil pressure of the fork to adjust compression damping according to the speed and distance of fork travel. And the unit was activated electrically whenever the front brakes were applied for instantaneous anti-dive effect.

BAC *(Balanced Actuation Calipers)*
These dual-piston calipers differed from others at the time by the pistons being of different diameters. This helped to balance the braking force and spread it evenly across the pad for less grab, better feel, and longer service life.

The ZX-10's only real failings were an extremely annoying mid-range flat spot and over-heavy steering

Much of the GPX750R's remaining technical specification followed that already pioneered on the earlier liquid-cooled four-cylinder models. Although an excellent bike in its own right, it was destined never to command the attention gained by the 600, 900 and 1000 models. In addition, as an out-and-out sportster its high speed handling was rather suspect compared to its Honda, Suzuki and Yamaha rivals, but against this it was probably a superior all-round motorcycle with good touring abilities.

The GPZ1000RX was replaced in the 1988 model year by the ZX-10. With 136 bhp, increased torque and higher maximum speed (electronically tested at a shade under 170 mph) the ZX-10 took over the laurels as the 'world's fastest production roadster'.

The 1000RX engine had been carefully improved, rather than given a complete redesign. Engine component weight was cut wherever possible, amounting to an overall reduction of 4 kg (9 lb). The cylinder head was completely new: it featured an individual-rocker valve train which reduced reciprocating mass for increased redline and added reliability. The head also featured straighter, semi-downdraft inlet ports, which were hand-polished for maximum induction efficiency. A larger airbox, semi-downdraft carbs and an improved version of Kawasaki's high-velocity induction system complemented the polished ports. More compact combustion chambers and a higher 11:1 compression ratio helped to maximise combustion efficiency, while lighter pistons and a lightened crankshaft/balancer allowed a 500 rpm redline increase as well as improved reliability. Digital ignition and a lighter, more efficient exhaust system rounded out a combination of features that in total added up to a power output unmatched by any other production streetbike.

In the area of chassis design, Kawasaki at last bowed to fashion and produced a new 'E-box' frame of extruded, dual-box-section aluminium. It was 4.5 kg (10 lb) lighter than the 1000RX's steel/aluminium item and, Kawasaki said, also offered increased rigidity.

In combination with the lighter engine, the new chassis helped reduce overall dry weight to 222 kg (490 lb) – a full 16 kg (35 lb) lighter than the 1000RX.

The ZX-10's bodywork was 7 per cent more slippery than its predecessor. There was improved suspension, including stronger 41 mm front forks. The ZX-10 offered the first dual front floating discs mounted on a Kawasaki streetbike. The combination of new 17 and 18 inch cast aluminium wheels featured large hollow hubs to maximise rigidity while minimising unsprung weight. Although the wheels were larger in diameter, the tyres' lower profiles ensured that overall wheel/tyre diameter – and gyroscopic effect – was similar to the 1000RX's.

Attention to detail had also not been overlooked. The one-piece dual seat was longer, wider, and lower, making for an easier life for both rider and pillion passenger alike. There were new four-way clutch and brake lever adjusters, a centre stand, a side handle to ease stand-setting, a retractable passenger grab rail, a single holder for two helmets, and an easy-access, lockable storage compartment set into the left side of the tail section.

While undoubtedly fast, the ZX-10 was not an ultra-sporting bike in the same way as the Yamaha FZR1000 or Suzuki GSXR1100. What it did offer was a good combination of usable power output and two-up touring ability. Its only real failings were an extremely annoying flat spot in the mid-range and over-heavy steering.

Utilising technology developed by the factory race team in the 1988 World Endurance Championship, Kawasaki introduced the ZXR750. Physically, this was very much a machine in the mould of Honda's RC30. In standard production form it produced 106 bhp at 10,500 rpm. Not a lot more, it has to be said, than the GPX750R. In some ways this was not surprising considering the engines had much in common (including the same 68 × 51.5 mm bore and stroke). The biggest changes came in the cylinder head design, valve train, induction, exhaust, clutch and perhaps most of all the all-aluminium perimeter type frame, similar in configuration to the one fitted to the ZX-10.

The 1989 (H1) and 1990 (H2) models were very much a case of a sheep in wolf's clothing.

For 1991, Kawasaki rectified the situation by bringing in a greatly revised ZXR750 and a performance version suitable as a basis for Superbike racing, the ZXR750R. Whereas the original versions of the ZXR had been largely ineffective from the sporting viewpoint, the newcomers boasted zip aplenty.

Though in different states of tune, both the new bikes shared an all-new cylinder head, valve train, pistons, crank and oil system. The new design featured a new short-stroke (71 × 47.3 mm) configuration, while the cylinder head utilised an offside cam drive and a lightweight individual rocker train to allow an amazingly compact 20 degree included valve angle. There was a larger 8.7-litre airbox and a less restrictive four-into-one exhaust. Technical differences between the two engines that altered power characteristics included carburettors, cam timing, cam duration and compression ratios. There was an improved frame borrowed directly from the factory's F1 race bike; it also had the advantage of being 4 kg (9 lb) lighter. There was a new cartridge-type (upside down) front fork with 43 mm stanchions (41 mm on the ZXR750R); plus improvements to the rear suspension featuring a long, threaded top shock mount to enable riders to alter the ride height, which in turn changed the steering characteristics to suit individual rider preference. Other changes included larger 320 mm (previously 310 mm) semi-floating discs at the front, and wider rims and tyres and F1 works style bodywork – the ZXR750 has a dual stepped seat, the 'R' version a solo seat.

The new breed of ZXR has proved itself more than equal to the task both on the street and the race circuit. One is left wondering why Kawasaki didn't do the job correctly in the first place?

There is also a smaller ZXR, the 398 cc (57 × 39 mm) '400', this too was revamped for 1991 with engine and transmission improvements, a new frame and swinging arm, upside-down front forks, more powerful brakes and significantly less weight.

Except for the recently introduced ZXR750/400 and the long running GPZ900R, it was all-change for the 1990 season. Out went the GPZ/GPX 600 and the ZX-10, in their place came the all-new ZZ-R series in 600 and 1100 categories.

It was Kawasaki who had created the 'hot six hundred' class (with their top selling GPZ600R), but since then Honda and Yamaha, and to a much lesser extent Suzuki, had got their act together; the result being a hyper-competitive category, with all the 'big four' Japanese manufacturers striving to achieve race-track performance from a road going sports bike.

Such development is fine... up to a point. What tends to get overlooked in the search for the ultimate 'racer-replica' is real-world practicality. Enter the ZZ-R600. Compared with its rivals it was a generously proportioned machine that offered a good degree of comfort. That was not to say that it lacked any performance or handling capabilities.

This was Kawasaki's first 600 cc class machine to feature an aluminium mainframe. The engine, although based on the GPX, had been strengthened and given a power boost to almost 100 bhp.

Beefy 41 mm diameter front forks – the same size as those on the ZX-10 – helped to keep the front end in order. Rear suspension, braking, wheels and tyres were all upgraded to match the new level of performance.

The ZZ-R600 hasn't featured much in the Supersport 600 racing class, but it does offer the road-going rider a more comfortable and practical mount than the out-and-out race replicas on offer from Honda and Yamaha.

But the real star of the ZZ-R duo was the 1100.

It would have been easy for Kawasaki to have created a new flag ship by simply taking the top selling ZX-10, boring it out a bit, adding a couple of styling mods and selling it at a higher premium. Luckily this didn't happen.

The ZZ-R1100 (ZX-11 in the States) has an engine which features several major changes to boost torque and strength. The chassis was new; so too were the bodywork, wheels, suspension, tyres and the brakes. It also rectified the problems of mid-range flat spots and over-heavy steering which had afflicted the ZX-10.

The ZZ-R was designed very much as a Sports Tourer, but with class-leading performance combined with the ability to corner with the best of them.

Taking the ZX-10 engine as a starting point, the engineers punched out the bore size by 2 mm to 76 mm, giving a capacity of 1052 cc, added larger inlet and exhaust valves, introduced lighter concave-crown pistons and stronger con-rods with a larger diameter big-end bearing. At the same time the crank was given bigger diameter journals, and the clutch larger diameter driven plates – then the whole plot was canted forward 2 degrees. power output of the original 1990 C1 version was a massive 147 bhp giving a top whack of almost 180 mph (UK versions were restricted to 122 bhp to comply with the voluntary 125 ps limit). For 1993 the unrestricted version has been upped to a shattering 157 bhp/187 mph!

Race-car technology is used featuring a 'closed' fresh air system, with an induction vent just below the headlamp (two on 1993 models). Carb size is upped by 4 mm to 40 mm from the ZX-10's 36 mm units. Fuel economy is excellent – some 30 per cent better than Yamaha's FJ1200.

The chassis layout although similar to the ZX-10 has thicker walls as does the swinging arm, while the fork stanchion diameter is upped from 41 to 43 mm. The rear shock is charged with nitrogen for improved fade resistance.

To keep this very considerable level of performance potential in check, the ZZ-R1100 places a wide footprint on the tarmac; the V-rated radial tyres measure 120/70 – 17 at the front, 170/60 – 17 behind. The wheels they grace are a 3-spoke design with hollow hubs for lower unsprung weight.

Stopping power is equally impressive: four-piston calipers for each of the 310 mm floating front discs, plus a twin piston Balanced Actuation Caliper for the single rear disc.

During 1990 the world's motorcycling press voted the ZZ-R1100 'International Bike of the Year', as did readers of several journals from

Above and right
The ZZ-R1100 is seen by many observers as not only the quickest streetbike in production, but also the finest with its unique blend of performance, handling, sweet power delivery and everyday appeal

around the world, including the respected *Motor Cycle News*.

The following extract from the August 1990 issue of *What Bike?* is typical of what the press thought: 'The five stars say it all. And even if the ZZ-R1100 could clock no more than 150 mph the conclusion would be the same; its the best of the heavyweights by far. Having stunning top speed just adds to the image. In these days of increased specialisation, it comes as a surprise to find a machine which has both stunning performance and good road manners, which at once can shock its rider with both its speed and level of refinement. It may sound corny to say that I was sorry to give the bike back to Kawasaki, but I was and I haven't experienced that feeling for a long time. By my reckoning the ZZ-R1100 is easily the bike of 1990'.

There is also the not insignificant fact that the ZZ-R1100 is still the world's fastest production roadster...

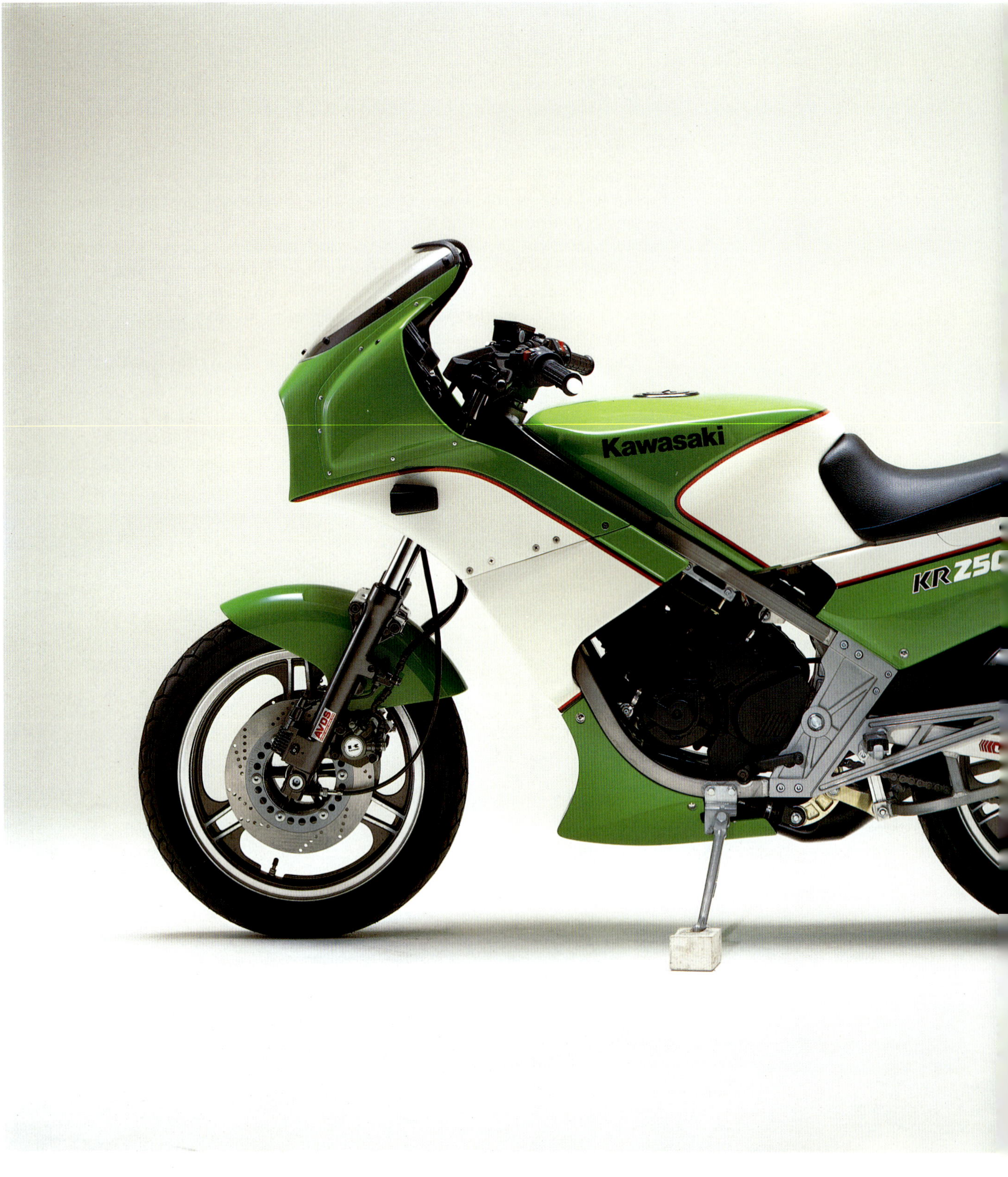
Kawasaki
AVDS

Two-stroke twins

Kawasaki's first quarter-litre two-stroke twin was the A1R racer of late 1966, closely followed by A1 Samurai, both of which are described in Chapter 1. With the advent of the three-cylinder S1/KH250 of the 1970s the company saw no need for a twin- cylinder two-stroke; instead it turned out a number of Z250 (Scorpion) based four-strokes. These lasted well into the 1980s and it wasn't until a totally new breed of high performance two-fifties arrived, led by the new Suzuki RG 250 Gamma that Kawasaki felt the need to act.

This resulted in the KR250 tandem twin of the mid-1980s based loosely around the world championship winners ridden by the likes of Kork Ballington and Anton Mang. The 249.1 cc (54 × 54.4 mm) model featured triple disc brakes, cast alloy wheels, Uni-Trak rear suspension, high hi-level exhaust system and three-quarter fairing; it was not imported into Britain.

Later in the decade the high performance 250-class became the most important on the domestic Japanese market. This resulted in the Yamaha TZR, Honda NSR, Suzuki RGV and Kawasaki KR1.

The first KR1, the B1, appeared in 1988. At its heart was an all-new 249 cc (56 × 50.6 mm) liquid-cooled parallel twin with crankcase reed valve induction. It ran on a compression ratio of 7.4:1 and gave 55 bhp.

The KR1 was the closest thing to a racebike that 'Big K' had ever put on to the street. Technical features included power valves, a cassette-type gearbox, high box-section aluminium frame and swinging arm, adjustable 41 mm fork, semi-floating 247 mm front discs with dual piston calipers, fully floating rear caliper, wide hollow-spoke wheels and low-profile radial tyres.

There were twin 28 mm Cresent-slide carbs, and to keep the engine's low-end and mid-range as wide as possible there was the new Kawasaki Integrated Power-valve system (KIPS). Controlled by a tiny computer and servomotors, the KIPS operated with maximum precision to knock out power right through to 11,500 rpm; but it is true to say that, like other

Making its debut in 1984, the KR250 was an in-line rotary and reed valve twin with a 6-speed gearbox. Other features of its specification included an aluminium alloy frame, air-ajustable and anti-dive equipped front fork and Uni-Trak rear suspension. It was never imported into the UK and was discontinued at the end of 1987

high-performance 250s, the best power is generated at the top end of the scale.

The factory claimed 131 mph and the standing start 200 metre ($1/8$ mile) in 12.5 seconds. Road testers raved and customers flocked to buy it. The KR1 was an immediate success, both on the street and in Production class racing events, which were soon to witness the birth of the now popular Supersport 400 class (open to 250 two-strokes or 400 cc four-strokes).

The first KR1 to be exported in any numbers was the B2 of 1989. This had a number of changes compared with the B1: new frame giving increased rigidity of the steering head and swinging arm pivot; the size of the box-sectioned aluminium beams, front cross beam, and seat rails was increased for extra strength; gussets had been added to strengthen the swinging arm; while bosses for steering damper and centre stand had been adopted. Performance and power output remained unchanged.

By the end of 1989 the KR1 was firmly established around the world, no more so than in Britain where at the final round of the national Supersport 400 Championship held at Brands Hatch, KR1s filled seven of the top ten positions, with John Reynolds setting a new class lap record of 50.8 seconds, on his way to clinching his fourth successive victory and being placed second overall in the series behind fellow Kawasaki rider Ian McConnachie.

If all of this seemed a hard act to follow, the pundits were caught napping as Kawasaki saw fit to launch an even more potent machine, the KR1S, for the 1990 model year.

The KR1 was never short on horsepower, but the KR1S had an extra 5 bhp, which was developed at the same rpm. This boost had been achieved by revised porting, a new exhaust system which included different expansion chambers and aluminium (rather than steel) bodied silencers, and new pistons. Another bonus was that the silencers and exhaust pipes were no longer one unit: in the event of damage, the silencers could be replaced separately. The KIPS was retained but with modified porting to help boost top end torque without sacrificing middle or bottom end output. The micro computer controlled ignition was also modified to provide a faster throttle response throughout the rpm band.

While the wall thickness of the twin spar aluminium frame remained unchanged, the cross section had been modified to provide increased rigidity – the smoother finish being another feature of the KR1S compared to its forerunner. The front cross brace was increased in size, and the cross section of the swinging arm was changed to increase torsional rigidity and welded gussets added to give extra strength. The length of the swinging arm was increased by 10 mm to allow fitment of a larger tyre (140/60 R18) and the outer diameter of the pivot spindle increased from 17 to 20 mm diameter.

Above right
The first of the new KR1 parallel twins, the B1, made its debut on the Japanese market in 1988. The B2 was the first version to be exported, a 1989 model of which is shown here in Britain that year. The punch of 55 bhp gave 125 mph and made the bike an instant hit for both street and track riders alike

Right
With 5 bhp more and a 130 mph potential, the 1990 KR1S was even better than the KR1. Team Green rider John Reynolds is seen in winning form during a Supersport 400 race at Cadwell Park in early 1990

Kawasaki
KIPS
KR-1

Arai
Kawasaki
RIZLA+
ROCK OIL
PIRELLI
MIRAGE
D.I.D
NGK
DUNLOPAD
GOODRIDGE
Arai
RIZLA+

KR18 Super Teen competitor Jonathan Peacock (centre) with his father Ray, mum Wendy, brother Timothy and the author's son Gary (far left) at Donington Park on 30 August 1992.. Gary later lost his life in a tragic racing accident.

Other changes included new style chain adjusters, shortened front forks, a nitrogen charged alloy-bodied rear shock with remote reservoir and revised spring rates, larger (300 mm) and thicker front discs, rear brake given a dual piston caliper, 5-spoke (in place of 3-spoke) alloy wheels, wider and lower profile radial tyres and revised cosmetics, including a new green/black colour option.

Throughout 1990 the KR1S was the machine to beat in the hotly contested SS400 racing class. In 1991 Suzuki uprated their V-twin RGV250 to become the 'M model', which at last proved the equal of the Kawasaki parallel twin. By 1992 the pendulum had swung even more Suzuki's way and the new breed of super-quick 400 four-cylinder four-strokes didn't help matters. Many are now asking the question: what now Mr Kawasaki? Whatever materialises, the KR1 and its successor the KR1S ruled the roost in 1988, 1989 and 1990, so their replacement would have a hard act to follow – three years at the top in the fiercely competitive quarter-litre sports class is a long time. How Kawasaki took up the challenge from Suzuki and elsewhere after 1993 will be the subject of a future Osprey Colour Classic.